Air Travel: How Safe Is It?

Air Travel: How Safe Is It?

Second Edition

Laurie Taylor
OBE, FRAeS

**Blackwell
Science**

© Laurie Taylor 1988, 1997
Blackwell Science Ltd
Editorial Offices:
Osney Mead, Oxford OX2 0EL
25 John Street, London WC1N 2BL
23 Ainslie Place, Edinburgh EH3 6AJ
350 Main Street, Malden
  MA 02148 5018, USA
54 University Street, Carlton
  Victoria 3053, Australia

Other Editorial Offices:

Blackwell Wissenschafts-Verlag GmbH
  Kurfürstendamm 57
  10707 Berlin, Germany

  Zehetnergasse 6
  A-1140 Wien
  Austria

First edition published 1988
Reissued in paperback 1991
Reprinted 1994, 1995
Second edition published 1997

Set in 10 on 13pt Times
by DP Photosetting, Aylesbury, Bucks
Printed and bound in Great Britain by
Hartnolls Ltd, Bodmin, Cornwall

The Blackwell Science logo is a trade mark of Blackwell
Science Ltd, registered at the United Kingdom Trade
Marks Registry

DISTRIBUTORS

  Marston Book Services Ltd
  PO Box 269
  Abingdon
  Oxon OX14 4YN
  (*Orders:* Tel: 01235 465500
         Fax: 01235 465555)

USA
  Blackwell Science, Inc.
  Commerce Place
  350 Main Street
  Malden, MA 02148 5018
  (*Orders:* Tel: 800 759 6102
              617 388 8250
         Fax: 617 388 8255)

Canada
  Copp Clark Professional
  200 Adelaide Street, West, 3rd Floor
  Toronto, Ontario M5H 1W7
  (*Orders:* Tel: 416 597-1616
              800 815-9417
         Fax: 416 597-1617)

Australia
  Blackwell Science Pty Ltd
  54 University Street
  Carlton, Victoria 3053
  (*Orders:* Tel: 03 9347 0300
         Fax: 03 9347 5001)

A catalogue record for this title is available
from the British Library

ISBN 0-632-04069-6

Library of Congress
Cataloging-in-Publication Data
Taylor, Laurie.
    Air Travel: How Safe Is It?/Laurie Taylor;
    Foreword by Sir Ross Stainton.–2nd ed.
        p.   cm.
        Includes index.
        ISBN 0-632-04069-6
1. Aeronautics–Safety Measures.
2. Aircraft Accidents.
3. Aeronautics, Commercial.
I. Title.
TL553.5.T34   1997
363.12'4–DC20   96-41630
                                       CIP

Contents

Foreword

Sir Ross Stainton, CBE
Formerly Chairman of British Airways

All forms of public transport need continuously watchful management, yet as everyone knows, accidents still happen.

Air transport, after the demise in 1930 of the lighter-than-air airships, depended upon the overcoming of gravity by means of static wings and forward thrust. No other means of transport faces the catastrophic problems which result from failure of lift or propulsion, leaving gravity in control. Primarily for this reason, air transport has from the very beginning accepted that safety is a prerequisite. Pilots, engineers, staff and management have to understand the safety implications of the machines in the air and their environment.

The law of the machines is poetically described by Rudyard Kipling:

'We can pull and push and lift and drive,
We can print and plough and weave and heat and light,
We can run and race and swim and fly and drive,
We can see and hear and count and read and write...
But remember please the Law by which we live,
We are not built to comprehend a lie,
We can neither love nor pity nor forgive.
If you make a slip in handling us, you die!'

In less than 80 years, transport aircraft have become one hundred times bigger in capacity, six times faster, capable of flying 60 times as far and at ten times the altitude. This spectacular development has not been without risks. What are the risks, and have they been acceptable? Laurie Taylor points to the answers, present and future. He has been a pilot of long and broad experience, particularly in international service and regulation. He is well qualified to review this broad field of air safety in its human, mechanical, environmental, regulatory and security aspects. His readers will be impressed by the breadth and complexity of this subject, basically one of deep public interest, involving governments and international agencies as well as the airlines themselves.

The challenge to transport managements and regulators is made even greater by the growing demand for travel as well as the introduction of very large vehicles, be they wide-bodied jets, drive-on ferries, super-tankers or trailer trucks, all of which have an added emphasis concerning the need for extra operational care and responsible management.

All air carriers are keenly aware of the demands of safety and security. Systems exist for the meticulous reporting and anlaysis of all accidents, and the lessons learned from them. There is continuous liaison between airline, aircraft builder and aviation authority. Such matters come under Board supervision. This book makes clear the dangers attending air transport and the measures involved in risk management. Governments and several international bodies share responsibility in such services as air traffic control and airway congestion, satellite navigation, meteorology, airport standards, and, indeed, security. Even so, a deliberate act of sabotage and a loophole in security, though hardly the fault of the airline, resulted in the Lockerbie disaster which contributed to the demise of Pan Am. It revealed the paramount need for concerted and harmonious international action.

Although the systems and policies of airlines and related agencies are well in tune with the public interest in this field, the same can hardly be said of other forms of transport, where the risks may be less obvious but are still serious. For example, the devastating results of recent ferry disasters and several large tanker accidents indicate a lack of attention at top level, not only to design weaknesses, but especially to the enforcement of sound operational practices and crew discipline. The same symptoms exist amongst bus and truck operators, although the consequences of accidents may be less spectacular.

Air Travel: How Safe Is It? is a valuable contribution to a wider awareness of this subject, and Laurie Taylor is to be congratulated on his assessment of the question and the answers.

Introduction

By the late 1990s, civil air transport had passed from a difficult beginning through several periods of extremely rapid growth and was close to becoming a mature industry.

Passengers carried on a yearly basis exceeded 1203 million in 1994 and with this increased number of lives at risk and important changes in regulation being introduced in many parts of the world, it is timely to examine the risks experienced in airline travel and to put forward views on where risk managers have succeeded and where they have failed.

I was a pilot for more than thirty years, initially with the RAF in World War II and then with the British Overseas Airways Corporation (BOAC) and its successor, British Airways. In addition to my experience gained in flying many types of aircraft, including the Boeing 747, over a worldwide route pattern, I was fortunate to gain other perspectives on civil aviation. There were years of technical work with the British Air Line Pilots' Association (BALPA), and a five year spell as its chairman. Five years on the Council of the United Kingdom's Air Registration Board was followed by membership of the Airworthiness Committee of the International Civil Aviation Organisation (ICAO) for five years, and a Flight Time Limitations Committee of the Civil Aviation Authority (CAA). Then there were two years as a Principal Vice President of the International Federation of Air Line Pilots' Associations (IFALPA), and ten years as Executive Secretary of that organisation. I was privileged to attend many meetings of ICAO and other international organisations in civil aviation, and all of that work fed a lifelong interest in air safety matters and made me the recipient of much safety information from worldwide sources.

For more than fifty years, I have seen excellent progress in the design, construction and operation of civil aircraft, with each new generation of the vehicle being safer than the previous one. The environment in which airline aircraft are operated has improved as regards air traffic control, airports, navigation facilities and systems, and weather forecasting.

Training provided for pilots has made great progress, and an understanding has been gained of the factors causing accidents. The chances of a passenger safely completing a flight, and of a pilot living to collect retirement benefits have steadily increased year by year.

However, lives have been unnecessarily lost, and I believe that more will be lost in the future unless greater efforts are made to improve the general level of safety to that achieved by the best operators of the best aircraft. In that matter, it is instructive to reflect that the Boeing Aircraft Corporation has produced statistics showing that 16% of operators of its aircraft were responsible for 80% of the total number of accidents suffered by those aircraft.

It seems that some organisations in the air transport industry are satisfied with the current level of air safety, and believe that there is no need to seek further improvements. This is an unwelcome change from past practice, where airlines consciously adopted safety standards and procedures exceeding those required by the regulatory authorities, and, if operators intend to barely meet regulations, this changed attitude may require that authorities abandon their former concept of the requirements being a safety net needed only on rare occasions.

All chapters of this book concern airline safety and include brief descriptions of:

- International bodies principally concerned with air safety
- Longer chapters on human factors in aircraft accidents and pilot training
- Accident investigation
- The aircraft vehicle
- The natural and operational environments
- Man-made problems such as hijacking and sabotage
- Other crimes and military interception of civil aircraft

Near the end comes a speculative look at safety problems to be faced in the near and middle future. Inevitably a few details appear in more than one chapter, but repetition is kept to the minimum necessary to make each chapter understandable without the need for using a system of cross references.

1: The Role of International Organisations

The civil air transport industry has grown at a greater rate than world trade, and is now dominant in international travel, with business travellers outnumbering tourists on some routes. However many airlines base plans for growth on international tourism. For many countries, tourism is the greatest earner of foreign currency, and even those countries that do not operate major airlines are acutely affected by the air travel industry.

The industry is capital intensive, with the associated aircraft, engine and equipment manufacturers employing highly qualified personnel, with a total output measured in billions of dollars. In the USA, the export of aircraft and equipment is the highest value item in its list of exports. In 1995, Boeing delivered $15.8 bn value of civil aircraft (notwithstanding a strike of production workers) and McDonnell Douglas $3.1 bn, compared with Airbus Industries' $9.6 bn. Outstanding orders for these three manufacturers were $60.6 bn, $9.2 bn and $51.8 bn respectively. In 1994, operating revenues of the scheduled airlines of ICAO contracting states (180 plus) were more than $247 bn, and with other expenditure by business and tourist passengers added, the economic significance of the civil air transport industry is obvious.

Air cargo flights carry an ever increasing percentage of the world's high value exported goods, and it seems probable that a historic pattern of growth in both the civil aviation operating and manufacturing industries will continue and perhaps accelerate (see Table 1.1).

With the number of passengers carried annually set to reach a total of two billion before the end of the century, the economic and safety performance of the civil air transport industry is vitally important to the world community. Most states in membership of the UN are also members of its specialized agency, the International Civil Aviation Organisation (ICAO), and most airlines are also members of the International Air Transport Association (IATA). A similar pattern is followed by national pilots' associations, air traffic controllers'

Table 1.1 World total revenue traffic, 1985–1994 (international and domestic services of airlines of the ICAO contracting states).

Year	Passengers carried	Passenger-km performed	Freight tonnes carried	Freight tonne-km performed	Mail tonne-km performed	Total tonne-km performed
	◄──────────────────────── Billions & (% change) ────────────────────────►					
1985	0.899 (6.0)	1.376 (7.0)	0.0137 (2.2)	39.84 (0.4)	4.4 (2.1)	167.69 (5.3)
1986	0.96 (6.8)	1.452 (6.2)	0.0147 (7.3)	43.19 (8.4)	4.54 (3.2)	178.8 (6.6)
1987	1.028 (7.1)	1.589 (9.4)	0.0161 (9.5)	48.30 (2.8)	4.7 (3.5)	196.46 (9.9)
1988	1.082 (5.3)	1.705 (7.3)	0.0172 (6.8)	53.27(10.2)	4.83 (2.8)	212.11 (8.0)
1989	1.109 (2.5)	1.774 (4.0)	0.0181 (5.2)	57.13 (7.2)	5.06 (4.8)	223.03 (5.1)
1990	1.165 (5.0)	1.894 (6.8)	0.0183 (1.1)	58.82 (3.0)	5.33 (5.3)	235.25 (5.5)
1991	1.135(–2.6)	1.844(–2.6)	0.0174(–4.9)	58.57(–0.4)	5.1 (–4.3)	230.61(–2.0)
1992	1.148 (1.1)	1.927 (4.5)	0.0173(–0.6)	62.61 (6.9)	5.12 (0.4)	241.95 (4.9)
1993	1.131(–0.6)	1.954 (1.4)	0.0175 (1.2)	67.48 (7.8)	5.23 (2.1)	250.08 (3.4)
1994	1.203 (5.4)	2.086 (6.8)	0.02 (14.3)	76.53(13.4)	5.47 (4.6)	271.5 (8.6)

associations and airport authorities, because all of them realize that an international industry cannot be served by paying undue attention to national boundaries and interests.

The International Civil Aviation Organisation (ICAO)

Many international organisations play an active role in promoting air safety, but the most important of all is the International Civil Aviation Organisation (ICAO), a specialist agency of the United Nations with more than 180 states as members. Its work directly affects all aspects of civil aviation, even in countries that are not members of the organisation.

Before World War II, attempts were made to devise an international framework for the development of civil aviation. However, none achieved more than a regional significance, except perhaps the International Air Convention of 1919 which was born in the aftermath of World War I and eventually achieved 38 ratifying states. The organisation formed as a consequence of the Convention, was based in Paris and was known as The International Commission for Air Navigation (ICAN).

World War II had a major effect on the development of the airplane, telescoping a quarter-century of development into only six years. The political and technical problems that were foreseen for post-war civil aviation therefore demanded early solutions in order to benefit and support a world at peace. Concerns were expressed with regard to legal and economic conflicts that might arise with peace-time flying across national frontiers, and the question of the installation and maintenance of air navigation facilities, often sited in remote places. In 1944, the government of the USA invited 55 allied and neutral states to meet in Chicago,and of these states 52 attended. For five weeks the delegates considered the problems of civil aviation, and the outcome was the (Chicago) Convention on Civil Aviation and the formation of ICAO.

Ninety-six articles of the Chicago Convention establish the privileges and restrictions of all contracting states, provide for the adoption of International Standards and Recommended Practices (SARPS) affecting air navigation, recommend the installation of navigation facilities by member states and suggest the facilitation of air transport by the reduction of customs and immigration formalities. The Convention accepts the principle that every state has complete and exclusive sovereignty over the airspace above its territory, and provides that no scheduled international air service may operate over or into the territory of a contracting state without its previous consent.

ICAO was formed in 1947 and in its 'Memorandum on ICAO' states:

'The aims and objectives of ICAO are to develop the principles and techniques of international air navigation and to foster the planning and development of international air transport so as to:

(a) ensure the safe and orderly growth of international civil aviation throughout the world;

(b) encourage the arts of aircraft design and operation for peaceful purposes;

(c) encourage the development of airways, airports, and air navigation facilities for international civil aviation;

(d) meet the needs of the peoples of the world for safe, regular, efficient and economical air transport;

(e) prevent economic waste caused by unreasonable competition;

(f) ensure that the rights of contracting states are fully respected and that every contracting state has a fair opportunity to operate international airlines;

(g) avoid discrimination between contracting states;

(h) promote safety of flight in international air navigation;

(i) promote generally the development of all aspects of international civil aeronautics.

'ICAO has a sovereign body, the Assembly, and a governing body, the Council. The Assembly meets at least once in every three years and is convened by the Council. Each contracting state has one vote and decisions are normally taken by a majority of votes cast. All the work of the organisation is reviewed at the Assembly's sessions, technical, economic, legal and technical assistance, and guidance is given to the other bodies of ICAO for their future work.

'The Council is a permanent body responsible to the Assembly and its members are elected for three year terms. It is arranged that adequate representation is given to states of major importance in air transport and that all geographic regions are represented.

Air navigation

'There is a continuous invisible link between aircraft and the ground stations and among ground stations themselves, and many ground stations are needed to ensure the safe and efficient operation of aircraft. To achieve harmonious functioning of all these ground facilities and services, international standardisation is necessary.

'To ensure safety, regularity and efficiency of international civil aviation operations, international standardisation is essential in all matters affecting them; in the operation of aircraft, aircraft airworthiness and the numerous facilities and services required in their support such as aerodromes, telecommunications, navigational aids, meteorology, air traffic services, search and rescue, aeronautical information services and aeronautical charts. A common understanding between the countries of the world on these matters is absolutely necessary.

'To achieve the highest practicable degree of uniformity, the ICAO Council adopts, and amends when necessary, International Standards and Recommended Practices, and approves procedures for the safety, regularity and efficiency of air navigation. The principal body concerned with their development is the ICAO Air Navigation Commission, composed of nineteen persons who have "suitable qualifications and experience in the science and practice of aeronautics". They are nominated by states and are appointed by the Council, to which body they report.

'Since ICAO came into being in 1947, the necessary standardisation has been achieved primarily by the creation, adoption and amendment by the Council, as Annexes to the (Chicago) Convention on International Aviation, of specifications known as International Standards and Recommended Practices (SARPS). The Standard is a specification the uniform application of which is necessary for the safety or regularity of international civil air navigation, while the recommended practice is one agreed to be desirable but not essential. There are nineteen sets of international SARPS and eighteen of them are within the technical field.

'The organisation also develops "Procedures for Air Navigation Services" PANS, and specifications known as Regional Supplementary Procedures for application in the Flight Information Regions to which they are relevant.'

Other activities of ICAO

Other ICAO activities relevant to air safety include: a computer based Accident/Incident Data Reporting (ADREP) program, studies related to aircraft noise and harmful emissions from aircraft engines, units of measurement, the carriage of dangerous goods by air, regional planning and implementation of regional plans (seven regional offices are maintained in Bangkok, Cairo, Dakar, Lima, Mexico City, Nairobi and Paris). In special circumstances the organisation arranges for joint financing of certain services and facilities where no individual nation can be charged with this responsibility. Air navigation services for North Atlantic flights are an example, where facilities based in Iceland and Greenland are jointly supported for the common good.

Legal matters

A permanent Legal Committee was established in 1947 and advises on matters concerning the interpretation of the Chicago Convention, and other matters referred to it. Eight legal conventions have resulted from this work, including three that refer to crimes affecting civil aviation, including hijacking.

Technical cooperation

For more than three decades the various international organisations of the United Nations have participated in a large scale multi-national effort to assist the technologically developing nations of the world to strengthen their economies. These efforts are conducted under the regular programs of some of the organisations, and under the United Nations Development Program (UNDP). While ICAO's technical cooperation is still a relatively small part of the UNDP, it has grown steadily and has achieved one of the best rates of implementation of all UN agencies.

Unlawful interference with international civil aviation

Faced with continuing occurrences of unlawful interference with international civil aviation, the international community has reacted in two different ways. In the legal field, the result obtained by ICAO has been three conventions signed in Tokyo, The Hague and Montreal, all with the aim of achieving universal acceptance and application of this legal structure so that offenders can and will be severely punished or extradited. In the technical field, the primary goal of the ICAO aviation security program is the prevention of unlawful acts, but it also deals with measures to be taken if preventative measures fail. To help states implement their own security programs, ICAO convenes informal regional seminars. It also publishes a manual which provides detailed procedures and guidance on aviation security. The organisation co-operates with other international organisations active in aviation security matters, such as Interpol, IATA, IFALPA and the Airport Council International (ACI).

The achievements of ICAO

The achievements of ICAO are great by any standard of measurement, and particularly in comparison with those of other specialist agencies of the United Nations. An international system of licensing and certification enables aircraft built in country A to be bought and certificated by country B and operated between countries C and D, without legal difficulties arising. Pilot licenses issued to ICAO standards are validated in other countries, with only minor variations in requirements being imposed. Air traffic control procedures are common throughout the world (although not necessarily performed to the same standard), a common system for the carriage of dangerous goods by air is implemented and navigation aids, airport runways, lighting systems and equipment are all standardised.

A remaining difficulty is that of implementing the ICAO Standards and Recommended Practices (SARPS) in countries lacking in technical and financial resources. It will require an increased application of technical cooperation programs over a period of many years to eliminate the worst problems.

In matters such as hijacking, ICAO has been confronted with problems that have political roots, with some member states being suspected of sponsoring or at least condoning the use of terrorist methods against civil aviation. Tremendous efforts made at ICAO Assemblies, and more importantly perhaps by Dr Kotaite the President of ICAO in his confidential missions to known problem spots, have greatly reduced the problem and the success of the three international Conventions against all aspects of what ICAO terms 'Unlawful Interference with Civil Aviation', also owes much to the personal efforts of Dr Kotaite.

Other air navigation problems of a political origin affecting civil aviation receive the attention of Dr Kotaite. Air traffic control problems in the Eastern Mediterranean caused by Turkish and Greek differences over Cyprus and other airspace over the Aegean Sea, have required missions to be undertaken by the President of ICAO. A similar problem in South East Asia and affecting air traffic over Vietnam, Laos and Kampuchea required a similar and successful mission. The shooting down of a South Korean airliner over the North Pacific, caused ICAO to mount a technical fact-finding mission in an attempt to diminish the long term consequences of the tragedy. All these activities reflect well on ICAO, its officers and secretariat.

In common with other organisations, ICAO has suffered when member states, including the USA, have not paid their dues or have paid them in arrears. By 1994, the organisation was owed $9.6 million and as total assessments for 1994 were only $48.4 million, the organisation has been forced to reduce activities, leave posts unfilled and cancel meetings. With so much effort being expended on Future Air Navigation Systems (FANS), other important activities were drastically curtailed in the late 1980s and early 1990s, and for long periods, work on accident investigation, dangerous goods and helicopters was virtually suspended. The seven regional offices were undermanned, and it seemed that the technical (air safety) work of ICAO was in permanent decline. However, with work on FANS almost complete, the organisation is taking steps to resume earlier activities, and there are encouraging signs that important air safety programs are being reinstated.

A defect ICAO shares with other international organisations, is that it can only recommend to sovereign member states that they adhere to Standards and Recommended Practices (SARPS) in important matters such as Security. No solution exists for the problems that arise when a state does not adhere to SARPS, but the record of ICAO in achieving its major goals is excellent, and far better than that of almost any other international organisation. Equally important is the fact that this achievement has been made without any charges arising of wasteful extravagance or financial mismanagement. A success story indeed!

The International Air Transport Association (IATA)

IATA was formed at the end of World War II for the purpose of promoting the interests of its member airlines, and it is properly described as a 'trade organisation'. Like other organisations of this kind, membership tends to fluctuate with economic circumstances, but most of the major airlines of the world participate in some of its activities. These include tariff and scheduling agreements and the transfer of revenues arising from interline transactions between participating members. At the end of 1995, membership of IATA stood at 233 airlines. Important activities include technical matters such as dangerous goods, security, training, flight operations, maintenance, fuel conservation and aircraft studies. Technical activities of IATA are based at Montreal and administration is at Geneva.

At ICAO, IATA is a recognised International Organisation, with permanent observer status on the Air Navigation Commission, a privilege of great importance. The only other organisation with the same privilege is the International Federation of Airline Pilots' Associations (IFALPA). Recognition by ICAO as International Organisations, enables IATA and IFALPA to participate in the work of almost all ICAO Committees, Panels and Secretariat Study Groups, but without having a vote. Representatives of IATA attend Regional Air Conferences, Divisional Meetings, Diplomatic Conferences and Assemblies convened by ICAO, again in an observer capacity, but with their views receiving careful attention. It is helped in its work at ICAO by many member airlines being government owned, leading to a probability that airline interests are understood by those states.

IATA has regular contacts with other international organisations that affect its activities in civil aviation, such as the World Meteorological Organisation, Universal Postal Union, Interpol, Airports Council International (ACI), International Federation of Airline Pilots' Associations (IFALPA), International Telecommunications Union and others.

IATA and air safety

It is probable that a passenger chooses to use a particular airline because of fares charged, reputation gained for standards of cabin service, punctuality and convenience of the aircraft schedule. For a brief period following an accident or hijacking, an airline may lose passengers to competitors, but in general terms most passengers tend to assume that all airlines are equally safe – perhaps because the public expects the authorities to regulate air safety.

Airlines have a collective interest in safety, and accept that their own particular accident rates will be reflected in the premium payments demanded by insurers. IATA shares the perceptions of its member airlines, but although it seeks to improve air safety, it also leads the airlines' continuous fight to reduce costs so as to remain in business. Proposals made by regulatory authorities, or by ICAO, to improve air safety, almost always lead to increased costs being borne by the air transport industry, and IATA therefore carefully scrutinises all such proposals and resists those that it believes to be 'unjustified' after taking into account the costs and benefits expected. Some judgements are difficult, and make the task of putting forward the IATA position at ICAO and safety meetings an uncomfortable one. Certainly they can lead to disagreements with ICAO, states and other organisations.

Interpol

Interpol's role in civil aviation safety matters is confined to the prevention of crime. However, as these crimes include sabotage of aircraft, bombings, hijackings, counterfeit aircraft parts and the smuggling of narcotics and illicit radioactive and nuclear materials (all of which pose a serious threat to human life), this role is one of great importance. The annual Congress of Interpol has a standard agenda item dealing with these matters, and ICAO, IATA and IFALPA are invited to the meetings with their representatives participating in the work of the appropriate committee.

The prevention of such crimes relies heavily on exchanges of information between the security forces of all states threatened by terrorists and criminals, with the threat being constantly reviewed and made the subject of secret assessments that are communicated on a 'need to know' basis. Forecasts made of upsurges of terrorist activity are usually accurate, although the particular target at risk may not be known. It is relatively common for particular airports and/or airlines to be placed on a state of alert because of information gained from Interpol or from another security intelligence source.

Airports Council International (ACI)

This organisation is directly involved in many aspects of aviation safety. Its members are responsible for ensuring that aerodromes under their control meet national and international requirements concerning runways, taxiways, approach and runway lighting systems, landing and navigation aids, fire and rescue services and some elements of air traffic control. In order to better perform that task, ACI participates in the work of some ICAO Panels, Committees and Study groups. ACI represents the 400 international airports and airport authorities who in 1993, handled more than 2 billion passengers and 35 million tonnes of cargo. It works through a system of committees dealing with such items as the environment, technical/safety and security.

In the past 20 years, airport security has become an added major responsibility, and the ACI participates in the work of the ICAO Panel dealing with aviation security, and with the work of the Committee on Unlawful Interference, when considered appropriate by that Committee.

International Federation of Air Traffic Controllers' Associations (IFATCA)

This organisation models its organisation upon that of IFALPA, and exists to promote the professional interests of air traffic controllers. A difficult occasion for IFATCA arose when the USA air traffic controllers went on strike in defiance of contractual obligations not to do so, bringing down the wrath of President Reagan, who dismissed all of the strikers. Sufficient controllers remained on duty to contend with a reduced flow of air traffic and the strike collapsed. The incident created stresses within IFATCA and serious disagreements between that organisation and IFALPA, particularly over whether safety was maintained in USA airspace when so many controllers were not at work. The pilot organisation held to the view that the reduced number of flights permitted in USA airspace was commensurate with the number of controllers on duty, and that there was no significant change in achieved levels of air safety. That opinion was supported by a safety audit made by the Flight Safety Foundation (FSF).

IFATCA's efforts are directed towards the improvement of airspace management, and it provides participating observers to work in studies made by ICAO in that field. The valuable contribution it makes to the improvement of air safety and regularity is acknowledged by all other international aviation organisations.

Flight Safety Foundation (FSF)

Although it is arguable that the USA based FSF is not a truly international body, it is sufficiently representative to qualify for inclusion in any account of flight safety activity. The FSF's high reputation for dissemination of safety information, and its educational and consultative activities earn it a great deal of respect. Seminars arranged by the FSF are attended by representatives of aircraft and equipment manufacturers, airlines, oil companies, state regulatory authorities, aerospace educational institutions, pilot associations, insurance underwriters, flight attendants' organisations, ICAO, IATA, IFALPA, and research institutions.

The officers and staff of the FSF have wide experience in aviation, and the organisation has been requested to carry out independent safety audits of some airlines and state organisations. One such safety audit was of the air traffic services provided by the Federal Aviation Administration (FAA) in USA airspace in the aftermath of the air traffic controllers' strike, after President Reagan had dismissed all of the striking controllers, and anxiety was being expressed about the consequences for air safety of that action.

The reputation that the FSF has, gains it contracts dealing with important safety issues including FAA contracts to develop a windshear training aid and to make a Flight Operational Quality Assurance (FOQA) study (an implementation of standardised flight data analysis programs) and work on a joint manufacturing industry/ICAO Controlled Flight Into Terrain (CFIT) study. The FSF sees itself as having the dual roles of safety advocate and ombudsman.

Society of Automotive Engineers (SAE)

Although this organisation was formed to coordinate professional opinion about problems faced by the infant USA automobile industry in the years after World War I, it quickly moved to other considerations and became a recognised authority that devised standards for such matters as the viscosity of lubricating oils, as shown by worldwide acceptance of designators such as SAE 10W–40. Since the end of World War II, the Society's interests have taken in all aspects of aircraft and spacecraft as well as of automobiles.

According to the SAE's Constitution, its purpose is 'to advance the arts, sciences, standards, and engineering practices connected with the design, construction, and utilisation of self-propelled mechanisms, prime movers, components thereof, and related equipment, to preserve and improve the quality of life.'

The Society forms committees to consider particular problems, and one that is of major importance to the civil air transport industry is designated as S7, – Flight Deck and Handling Qualities for Transport

Aircraft. Membership of the committee is by invitation, after expressing an interest. It includes representatives of manufacturers of civil and military transport aircraft, some manufacturers of aircraft equipment, major airlines, research centres, test pilots, USA ALPA, IFALPA and some aviation consultants. Innovative technological developments receive the committee's special attention, and fly-by-wire (or light) control systems, all-electric aircraft systems, new instrument displays and proposals for aircraft with reduced aerodynamic stability are typical examples.

International Airline Passenger Association (IAPA)

This consumer organisation was formed in the USA in 1960, to provide services for airline passengers and to put forward their views to airlines and authorities. IAPA has offices in four countries and more than 100 000 members who make more than eight million passenger flights per year. Views of members are obtained by means of surveys, and by members making direct contact with any of the service centers. The organisation maintains contact with IATA, IFALPA, FAA and other regulatory authorities, and with organisations representing cabin crews.

International Federation of Airline Pilots' Associations (IFALPA)

IFALPA was formed in 1948, and has more than 80 national pilots' associations in membership. Because of the apolitical nature of its constitution, leadership and administration, IFALPA is able to attract into membership pilot groups from all regions of the world and from countries of many different racial, religious and political backgrounds. The original member associations were mostly from North America and Europe, but membership in 1996 saw all six continents represented with the sizes of the national pilot groups varying from the 40 000 of the USA to the seven of Uganda. Most European pilot groups are members, and the gaps in membership are mainly from countries where socio/political systems do not concede the right of association to their pilots, or other citizens.

As in similar self-funded organisations, membership fluctuates with changes in economic and social conditions, with some pilot associations disappearing when the air transport of a particular country becomes a function of the military forces, or when democratic processes are set aside. The Federation accepts only one national pilot group per country into membership, and does not accept any form of sponsorship, subsidy or corporate membership. Elected leaders are all practicing pilots, with the salaried permanent secretariat being professionally qualified. Only airline pilots represent the Federation at external meetings, so ensuring that the views expressed at meetings of ICAO and other organisations are truly representative of professional pilot opinion.

At ICAO, professional airline pilots represent the Federation in the work of 28 Committees, Panels and Secretariat Study Groups, and a number of its representatives have additional professional qualifications that are of value in their chosen speciality subject. The Federation's 'permanent observer' participates in the work of the Air Navigation Commission and it is represented at almost all Divisional, Diplomatic and Regional meetings convened by ICAO.

The President of ICAO, Dr Kotaite, when opening the 40th Conference of IFALPA, stated:

'This conference provides another opportunity for fostering the valuable professional contact and mutual understanding that is absolutely essential to the success of our cooperative efforts in furthering the safety and reliability of air transport. IFALPA has made tremendous contributions to the work of ICAO, the interests of everyone in the aviation community and the public at large are well served by your direct involvement in the ICAO process...

'In closing, I would like to reiterate the importance that ICAO attaches to the support and assistance of IFALPA. The amount of time and effort contributed by your representatives, their technical knowledge and expertise, and their professional integrity and dedication reflects highly on your organisation and serves humanity well. In return, I can only say that we in ICAO have an absolute need to remain in contact with the flight operations experts of civil aviation. We need to be kept aware of, and up to date with, all significant problems that arise. We look forward to your continued cooperation over the next 40 years...'

The contribution best made by IFALPA at meetings of ICAO and other international organisations, is to bring to the attention of other experts, a first hand, current and professional opinion on the operational problems encountered around the world, and to comment upon the likely efficacy of proposed solutions. The Federation excels at defining operational objectives and as it does not care if the technology, hardware and procedures to meet those objectives originate in Russia or the USA, its objectivity is much appreciated.

Less well appreciated is the Federation's occasional lack of understanding of the many demands made on the budgets of developing states, some of whom rightly place a higher priority on providing a supply of clean water for their citizens, than they do to installing a second Instrument Landing System (ILS) at a national airport.

IFALPA and air safety

Some of IFALPA's major successes in solving particular safety problems have occurred because of its willingness to take direct action to apply pressures, and to publicise its assessments of what is required to be done to meet particular circumstances. States and airport authorities do not welcome being 'awarded' an IFALPA 'Black Star' classification for what the Federation describes as 'critically deficient' airspace or aerodromes. The resultant publicity affects public opinion and the tourist trade that is so important to many national economies.

Internationally, IFALPA has used its power with discretion, and only with the support of public opinion. On several occasions it has organised, or threatened to organise, international stoppages of civil air operations in order to persuade international organisations, or particular states, to take action to end or reduce particular threats to air safety. These actions have been taken in cases of hijacking and military interceptions of civil aircraft where passengers' lives have been lost or put at risk.

2: Human Factors

Although great progress has been made in the design, manufacture and operation of civil aircraft in the past fifty years as evidenced by the steadily decreasing rate of accidents, more needs to be done to provide a better match between the aircraft vehicle, environment and crew members who operate aircraft. Aircraft manufacturers make great efforts to improve aircraft structures, systems, and engines, and the responsible authorities make similar improvements in weather forecasting, air traffic control and aerodromes. The pilot, however, remains much the same as he or she was in the earliest days of aviation. They have the same number of senses (ten in all), comprising: visual (sight), auditory (hearing), tactile (touch), kinesthetic (muscle sense), balance, olfactory (smell), taste, cold, heat, and pain. The first five are of critical importance in flying and are given great attention in the selection and training of pilots.

Normal ($1\,g$) flight in visual flight conditions places only small demands on pilots and their senses, but abnormal flight producing high g forces in instrument flight conditions can produce in-flight spatial disorientation. This occurs when a person's ability to perceive motion, position and attitude in relation to the surrounding environment by use of visual, vestibular (inner ear), muscular and skin receptors is disturbed. Spatial orientation mechanisms can encounter interpretation problems when visual references are lost and an illusory environment exists. These conditions exist when accelerations (e.g. spinning motions) occur and under certain other situations, including the influence of diseases, narcotics and alcohol. A number of military airplanes have been lost due to in-flight spatial disorientation, and the phenomena has been a causal factor in airline accidents when a pilot has lost control of an airplane.

Readers of official reports of air accidents are struck by the inexplicable nature of some of the mistakes apparently made by pilots who were very experienced, and who had previously held an accident-free record. In 1989, a B737 of China Air Lines crashed in the central mountains of Taiwan, ten minutes after take-off, as a result of turning left instead of right. The initial climb should have been over the sea, but on this occasion an experienced pilot flew west into the Chiashan mountains causing the death of all the passengers and crew. It is possible to speculate that the pilot was incapacitated, or simply made a mistake, but it is difficult to understand why other members of the crew did not detect the error and point it out to the captain, or physically intervene to prevent the disaster.

Pilots express concern when required to fly several types of airplane concurrently, although they accept that manufacturers are usually successful when they set out to provide near-identical flying qualities for a family of airplanes, e.g. the Boeing 757/B767 and the Airbus A320/A330/A340. They are most concerned when there are important differences beween airplanes that may appear to be similar. W.P. Monan, an aviation research consultant to NASA's Air Safety Reporting System (ASRS), has stated that non-standardised cockpit configurations were causally related to more than 150 'events' that threatened air safety, including 110 instances of flying at the wrong altitude, the so-called 'altitude busts'. Pilots making these reports are quoted as stating, 'We are operating twelve airplanes and have ten different cockpit configurations', and, 'It is not uncommon to fly three different models in a single day', and, 'We may go for months between flight segments on the model X'. Airlines make significant savings in costs when pilots operate several different types of airplanes, but regulatory authorities should be alert to the operational and safety consequences of authorising multi-type operations, particularly when the differences are subtle and likely to cause errors. The problem is greatest when operators build an aircraft fleet from different sources, and particularly in the so-called 'developing world', where regulatory authorities may not have the expertise and experience to monitor airline operations. In all cases it should be required that pilots remain 'current' on all types they are qualified to fly, i.e. they do not go for long periods without flying a type on which they are qualified.

There have been many instances where the design of aircraft has been improved because of lessons learned from the study of accidents and the recognition of mistakes, but this is no longer an acceptable means of making progress in an industry that regards itself as being in the forefront of scientific and technical progress.

Research establishments, universities and other organisations are currently active in the study of human factors, and the national authorities who are responsible for regulating the airworthiness of aircraft, the operation of air transport organisations and the licensing of pilots, air traffic controllers and engineering organisations, accept much of their advice as becomes available. Some suggestions gain acceptance from the International Civil Aviation Organisation (ICAO), and the 180 plus states who are members are expected to apply the agreed Standards and Recommended Practices (SARPS) of that organisation. Human factors affect the SARPS that apply to medical and professional standards applied to personnel who are licensed to carry out safety-related duties. Typical examples are the visual acuity and hearing standards required of an applicant for a professional pilot license.

From these few examples, it can be seen that some progress has been made towards the recognition of human factor problems, but many more remain unsolved or even unrecognised. They range from the variable performance of pilots under stress, adverse influences of alcohol or drugs, the ageing process, the effects of chronic fatigue and the stress and psychological pressures created by family or social problems or by employers, or by a basic and serious incompatibility between members of a crew. Technological changes making possible reduced crew complements, all-weather operations and supersonic/space flights, are expected to bring about significant changes in the way in which aircraft are operated, with consequential and new challenges arising in human factors matters.

A particular human factors accident (Mount Erebus)

In November 1979, a New Zealand registered DC10 aircraft left Auckland Airport on one of a series of passenger flights arranged to take place over the ice cap of Antarctica for sightseeing purposes. Passengers and crews from earlier flights had reported on the magnificent scenery of ocean, mountains and glaciers, and the airplane was carrying a complement of 257 passengers and crew. The flight ended in total disaster when it collided with the slopes of Mount Erebus. A subsequent accident inquiry and a later Royal Commission, disagreed sharply on who should bear the primary responsibility and blame for the disaster, but general agreement exists that there were two main causes for the accident. In order of occurrence, they were firstly the changes made to the route provided for the aircraft's automatic navigation system, made by the navigation section of the airline without the crew being informed of the change (the new route took the airplane directly over Mount Erebus), and secondly the crew flying into the side of the mountain without being aware of the danger. There were therefore two human errors, one by the ground personnel and one by the flight crew. It is virtually certain that the aircraft was in a controlled descent at the time of the disaster, with the pilots convinced that they were flying in 'visual conditions'. In these conditions pilots would expect to be able to avoid collision with another airplane or with the ground by means of the 'see and avoid' technique. The visual phenomena prevailing at the time were those known to persons experienced in Polar conditions as a 'white-out'. In these conditions the horizon cannot be perceived, snow covered ground and grey skies are merged, and a pilot would be deceived into believing that he or she was at a safe height above the ground. It has been shown that in a 'white-out' condition, a dark object is visible for many miles, while a snow covered object, even a mountain near to the observer, would be invisible.

Other visual illusions

When a pilot is making a visual approach to land, he or she places great reliance on a perceived perspective of the runway during the later stages of an approach. Flight characteristics of all fixed-wing aircraft, except those especially designed for short take-off and landing, require that flight path approach angles be as close as possible to $3°$ above the horizontal, and in visually good conditions an experienced pilot can detect errors of about $\frac{1}{4}°$ without reference to flight instruments. There are however a number of possible adverse conditions that make it unwise for a pilot to place undue reliance on that aquired ability.

Refraction caused by water on the windshield is one such condition, and has caused landing accidents by convincing the pilot that the aircraft was higher (at a steeper flight path angle) than was the actual case, causing the pilot to land the aircraft short of the runway. Another common cause of a false judgment being made concerning the flight path angle, is when the pilot's eye is not at the 'reference' position because the vertical adjustment of the pilot's seat is incorrect. In addition to providing a false perspective, too low a seat position causes the visual ground segment to be greatly reduced. When an aircraft is in a landing configuration at 100 feet above the ground with a surface visibility of 1200 feet, and the pilot's eye is at the reference position, it should be possible to see approximately 600 feet of the runway. Should the seat be only one inch too low, the pilot will see only 340 feet of the runway, and may as a result decide that it is not possible to make a safe landing. Too high a seat position also distorts perspective and again will result in a misjudged approach and landing.

There are other adverse conditions that cause landing accidents, one being misleading visual effects caused when an air temperature inversion occurs, frequently at dawn, causing the boundary between the lower (cooler) air, and the higher (warmer) air to act as a mirror and to cause serious refraction of all the pilot's visual 'cues'. A number of heavy landing incidents have been caused by this particular visual illusion. Another visual illusion that pilots are warned about includes a false perspective caused when an airport is situated on sloping terrain, and infrequent exposure to any of these illusions can lead to a failure to recognise them when they occur.

Further causes of optical illusions are fog, snow, night, and irregular terrain on the approaches to a runway, and the known existence of these hazards is a major reason for the installation of non-visual and visual landing aids at airports. These range from Instrument Landing Systems (ILS) that project a radio-magnetic beam aligned with the runway and down which an aircraft flies to a landing, through to Approach Light Systems and Visual Approach Slope Indicators (VASI), all of which have as a prime objective the removal of reliance on the real world visual 'cues' that must be used by pilots in the absence of such landing aids.

The known existence of all these problems of imperfect human perception was a major factor in the development of automatic landing systems that place little reliance on what is seen or not seen from the flight decks of aircraft, save for the monitoring of the performance of the aircraft systems by the flight crews. Widespread use of automatic landing systems in all weather conditions has made landings safer, and by early 1996 there had been no fatal accidents attributed to these operations although many thousands of landings have been made in adverse non-visual flight conditions.

Errors in reading flight instruments

In a series of laboratory tests conducted in the USA, it was demonstrated that when people were invited to read instruments with five different dial formats, vertical, horizontal, semi-circular, circular, and 'open-window' (albeit with the same scale length, pointer width, graduations and design of numerals), marked differences arose in the number of mistakes made (see Figure 2.1). In each case the same value to be read was provided with no interpolation being necessary. The errors made were as follows:

vertical	35.5%
horizontal	27.5%
semi-circular	16.6%
circular	10.9%
open-window	0.5%

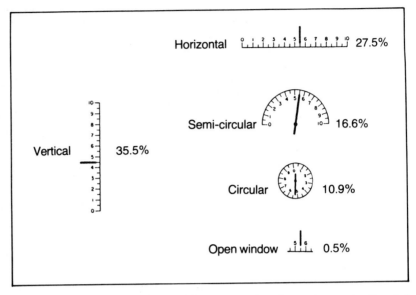

Figure 2.1 The errors committed by 60 people who took part in the dial experiment (*Courtesy of RAeS Aerospace*).

With such large discrepancies in performance, it is clear that great care must be taken to ensure that the best choice is made in the design of flight instruments, if serious and perhaps fatal, instrument reading errors are not to be made by pilots.

Some states permit the use of three-pointer and drum-pointer altimeters in civil aircraft, although both types have been shown to be conducive to dangerous reading errors and to have been a factor in fatal aircraft accidents. Pilots are particularly likely to misread them by 1000 feet when aircraft workload is high during landing approaches, and when these circumstances are combined with reduced visibility, an accident may result. Although improved types of altimeter are fitted to new types of aircraft, the old 'killer' types continue to be used in hundreds of aircraft that have years of useful life remaining. A test carefully carried out in 1969 (see Figures 2.2 and 2.3) showed that the difference between the 'best' (counter-pointer) and 'worst' (three-pointer) altimeters were:

	Counter-pointer	Three-pointer
Average reading times (seconds)	1.34	3.90
Reading errors (%)	0	20

In a fatal accident in Manitoba, Canada in 1994, it was found that the pilots were using two very different altimeters, the captain a drum-

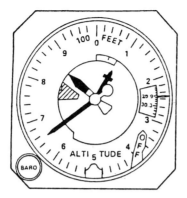

Figure 2.2 Three-pointer altimeter.

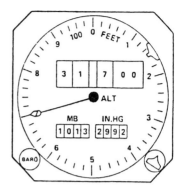

Figure 2.3 Counter-pointer altimeter.

pointer type and the co-pilot a three-pointer type. It is not surprising that a major finding of the accident inquiry was that 'the flight crew lost altitude awareness during the localiser back course approach...'

When all airplanes are fitted with the latest type of electronic flight instruments systems (EFIS), this long standing threat to air safety may finally be removed, as the systems can be programed to show any desired (optimum) display, without the expense of installing new instruments.

Units of measurement (pressure altimeters)

An additional and important factor in the problem of ensuring that a pilot always has accurate information regarding the height of the aircraft, lies in the failure of ICAO to ensure a common system of units for measurement in the vertical plane. The two systems in use are based on the meter or the foot respectively, and although the ICAO recommendation is for eventual adoption of the metric system, it will be many years, if ever, before standardisation is achieved. Until that time comes, pilots find themselves using altimeters calibrated in feet and being required to report their altitude in metres – or vice versa – according to the regulations of the state over which they are flying. The adopted palliative – unsatisfactory from a human factors standpoint – is to use a conversion card to convert a reading of the instrument reading to the other unit of measurement. This practice introduces a high risk of error and an undesirable slow-down in communications between air traffic controllers and pilots.

Information transfer

Studies of human error suggest that problems associated with the process of information transfer account for a large proportion of accidents and incidents, and that what might appear to be a bad decision may have been caused by a failure to transfer information correctly. Accidents or incidents may be caused by a failure to pass information to the person who needs it, or by passing misleading or false information, or by the recipient misunderstanding information correctly passed.

Pilots obtain information from the 'hardware' that surrounds them on the flight deck, such as instruments and displays, and from 'software' such as charts, manuals, and check-lists. They also receive information from 'live-ware', such as other crew members and air traffic controllers, and they receive information from the environment in which the aircraft is operated. Weather, turbulence, horizon (if it is visible), cloudscapes, light, dark, the ground (if it is visible) and aerodrome lighting systems, are only some examples of environmental information used by a pilot.

Radio procedures

Transfer of information between air traffic controllers and pilots is a potent source of error, and this problem required the attention of an ICAO Study Group with membership drawn from representatives of states, airlines, pilots, and controllers. The study group reviewed cases where it had been established that serious errors or misunderstandings had occurred, causing accidents or incidents.

One of the best known is the disaster that resulted from the collision of two Boeing 747 aircraft on the runway at Teneriffe on 27 March 1977. Fog was a major factor in the accident, and the pilot of one aircraft started his take-off run while the runway was still obstructed, because he believed that he had been cleared for take-off although a transcript of the radio transmissions shows that he had not. He also used the misleading phrase 'We are now at take-off', causing the air traffic controller to believe that the airplane was at the take-off point, *not* taking-off.

A NASA researcher reported to the Flight Safety Foundation (FSF) in 1995, that he had identified more than 200 communication related aviation incidents, some of which had resulted in disastrous accidents. Many of these incidents were linguistic-based, perhaps exacerbated by non-linguistic factors such as distractions, fatigue, impatience, obstinacy, frivolity, or conflict. Out of more than 6000 reports submitted by pilots and controllers to the Aviation Safety Reporting System (ASRS), there were 529 reported incidents classified as representing 'ambiguous phraseology'. Problems arise from *homophony*, the occurrence of different words that sound almost alike, such as *left* and *west*, or exactly alike such as *to* and *two*. The latter misunderstanding actually led to a fatal accident at a southeast Asian airport, when air traffic control cleared the aircraft to descend 'two four zero zero' and the pilot read back the clearance as 'OK. Four zero zero' and descended to four hundred feet! In another case a captain heard his copilot say 'cleared to seven'. He began a descent to 7000 feet when the copilot advised him that the correct altitude was 10 000 feet and the clearance was in fact, 'cleared two seven' meaning land on runway 27. It would be possible to provide many more instances of such misunderstandings.

Another potentially dangerous source of misunderstandings arises when two aircraft have similar radio callsigns. It is common practice for flight numbers to be a major part of an aircraft's radio callsign, and for two aircraft to be flying to the same airport at the same time and be using the same flight number, albeit with a different prefix. In these circumstances, there are many instances of controllers confusing two aircraft, or of the pilot of one aircraft acting on an air traffic control 'clearance' intended for another. In the crowded skies that surround busy airports the potential for disaster is obvious, and there have been many near mid-air collisions as a result.

The use of abbreviations of radio callsigns creates problems, with a main reason being an apparent lack of radio discipline by pilots and controllers, caused mainly by the severe congestion found on radio frequencies in busy airspace. A frequently experienced source of difficulty in busy airspace arises when radio transmissions are blocked by other simultaneous transmissions, causing serious misunderstandings between pilots and controllers. A technical solution for this problem is available, a device weighing only one pound that detects when a channel is active and prevents a transmitter being used until the channel is free.

The use of a data link between ground and air, with hard copy printouts available if required, is expected to bring some relief of this problem, although it seems certain that the more flexible and direct VHF voice transmissions will continue to be used for 'tactical' negotiations between pilots and controllers.

Flight instruments and procedures

Altimeters provide useful information only if they are referenced to the existing atmospheric pressure, for they measure pressure and not altitude. A subscale is set by the pilot with the calibration of that subscale being in millibars or inches of mercury. If a pilot sets the atmospheric pressure prevailing at local ground level, the indicated altitude will be above ground level, and if the sea level pressure is used as the reference, the indicated altitude will be above sea level. It is therefore necessary for the air traffic controller to pass the prevailing barometric pressure to the pilot before a landing approach is commenced. If the wrong pressure is set on the subscale of the altimeter by the pilot, a false altitude will be indicated.

Some years ago the pilot of a British operated Comet aircraft landing at Nairobi airport, which is more than 5000 feet above sea level, was passed the 'altimeter setting' in millibars. When the setting was used by the pilot to set the subscale of the altimeter, he transposed the digits, setting 938 instead of 839 millibars and caused the altimeter to over-read by almost 3000 feet. It was a near miracle that when the aircraft landing gear hit the ground miles before the runway, in poor visibility, it was in one of the few level places in the Nairobi Game Park. The aircraft bounced into the air and the pilot was able to recover the situation, climb away and land safely at the airport. The aircraft was not damaged but the pilot's career was!

The inevitability of humans making a mistake in transposing digits in a long number is well known to all of us who have used telephones, and it is remarkable that procedures were not devised earlier to prevent such simple but serious mistakes arising in aircraft. Monitoring procedures are the only way to avoid such simple but critically important mistakes arising, and all flight crew members are assigned this task.

Another accident similar to the Nairobi one, but resulting in a heavy loss of life, occurred when the pilot of a USA registered DC8 aircraft landing at Calcutta Airport was passed the 'altimeter setting' in millibars, but wrongly assumed that it was an abbreviated form of a setting in inches of mercury. He heard the air traffic controller give the setting as 992 millibars and assumed that the setting was 29.92 inches of mercury. That mistake caused the altimeter to over-read by approximately 600 feet and the aircraft crashed in poor visibility with a heavy loss of life. Only when a single system of units of measurement is adopted globally, will this type of error be ended.

Language

English is the language most used in civil aviation, perhaps because English speaking nations have tended to dominate the design, manufacture, sales, and operation of civil aircraft. This dominance, although challenged, seems likely to continue and the English language itself is a feature of any study of human factors. In the *ICAO Journal* of January 1996, an example is provided of a maintenance error caused by a less-than-optimum choice of words when a certain maintenance procedure was 'proscribed' (i.e. prohibited) in a service bulletin. The technician reading this concluded that the procedure was 'prescribed' and proceeded to carry out the forbidden action. There is therefore a strong case for simplified English to be used so that the words used mean the same thing to every reader.

When the prime contractor for the USA's Apollo space program carried out an experiment on how mechanics understood instructions, the following results were achieved:

	Accuracy
Combined written and oral instructions	77%
Oral instructions	61%
Written instructions	49%
Bulletins	37%
'Grapevine'	36%

It is clear therefore that combined methods produce the best results, and the lesson has been accepted in the training of pilots who receive information from manuals, lectures, teaching machines, and 'hands on' training sessions carried out on procedures trainers, flight simulators and aircraft. Other research programs have shown that the written word is best understood if short sentences of less than seventeen words are used. A simple style of writing and the avoidance of the use of conjunctions is recommended.

The internationally accepted language for air traffic control is English, with the national language of the state providing the air traffic control service being the only other acceptable alternative. Air traffic controllers are required to reply in English if pilots use that language in their first radio transmission. Because the majority of air traffic is of an international nature, except perhaps in the USA, Russia and China, most communications in civil aviation are in the English language. Pilots who fly international routes necessarily learn to speak English, and their task is made easier by the adoption of internationally agreed standard phraseology.

Difficulties can arise if a pilot is not fluent in the English language and flies to a very busy airport such as Chicago's O'Hare, where a high traffic flow rate requires controllers to speak more quickly than is usual at other airports. When a fast rate of speech is combined with a controller's use of idiomatic phrases, slang, and strong regional accents unfamiliar to the foreign pilot, serious difficulties arise and there is always a risk that a misunderstanding may lead to an incident or accident.

Some airlines of non-English speaking countries require flight crews to use the English language at all times when in flight, and their manuals and check lists are produced in that language. That particular solution to the problem has been rejected by other airlines for reasons of national pride, and the possibility of serious misunderstandings of spoken communications between pilots, and between pilots and controllers therefore continues to exist.

Aircraft warning systems

Aircraft warning systems are used to provide information to flight crews, and the design and performance of some is a source of controversy between system designers and pilots. The landing gear warning system is a typical example.

Landing gear warning systems are intended to warn the pilot that it is unsafe to land when he or she reduces power by closing the throttles without having first lowered the landing gear, and in these circumstances a red light is displayed and a warning horn sounds. There are, however, many occasions when a pilot reduces power for the purpose of reducing speed or to descend, without intending to land. This causes the warning system to operate as designed, creating an unwanted distraction with the warning horn being so loud that it prevents almost all communication and thought. To contend with this unwanted situation, a sub-system is provided to enable the pilot to silence the warning horn. The result is that because the warning horn is more often unwanted than wanted, flight crews are conditioned to silence the warning horn almost every time it sounds. There have therefore been occasions when crews have ignored a genuine warning with an inevitable result – an example of the design objective of a warning system not being met.

Many warning systems are installed in large civil aircraft, and pilots complain about the possibilities of confusion arising from that practice. There are as many as ten different aural warning systems including bells, horns, intermittent horns, clackers, chimes, plus recorded or synthetically generated voices, all intended to convey significantly different information.

When problems associated with the use of a hundred or more flight deck warning lights of red, amber, green, blue, and white colours, were added to the possible confusion caused by multiple auditory warnings, it was obvious that a complete and thorough re-design of aircraft warning systems was needed. Later types of aircraft are improved in this respect and are fitted with centralised warning and monitoring systems that apply logic in determining what warnings should be displayed to pilots in the many different circumstances that may arise. This reduces the possibility of unwanted confusion and stress arising from badly designed warning systems.

Human stress

All humans experience stress at various times, and it is widely held that minor stresses may be a 'good' way of motivating a person and ensuring alertness. Interest shown by psychologists in mental stresses experienced by pilots, arises from a probability that stress has been a causal factor in some aircraft incidents and accidents.

The 'hurry-up' syndrome is a stressmaker, and occurs when pilot performance is degraded by a perceived or actual need to hurry tasks or duties. These time-related pressures include: the need to expedite taxi for take-off, or to meet a restriction in air traffic control clearance time, to keep on schedule when delays have occurred because of maintenance or weather, or to avoid exceeding flight and duty-time regulations. Such pressures create stress and degrade human performance. A 1993 study of 128 citations from NASA's Aviation Safety Reporting System (ASRS), showed 60% were errors of commission with required tasks accomplished incorrectly or tasks executed that were not required, producing unexpected and undesirable results. Also, 30% were errors of omission with some element of a required task not accomplished.

Pilot associations recognise that occupational stress is a particular threat to pilot careers and have identified the time constraints and associated limitations on the control of situations as the common stress factors in many cases. Further identified stress factors involve: consideration of the supervisory and legal environment in which they work requiring them to be subject to up to seven checks/tests on an annual basis, the possibility of involvement in accidents with the inevitable investigations and inquiries and a consequential risk of assessment of blame. Another significant factor in producing stress is the occurrence of life event changes, such as family bereavement or a divorce, and research suggests that many pilots exhibiting symptoms of mental illnesss or who have been involved in an accident have recently undergone a life change of that kind.

Crew schedulers of major airlines believe that they are sometimes the first to see a potential case of mental illness arising, when crew members request that they not be scheduled to fly with a particular individual. Pilot associations believe that excessive stress may be a causal factor in many incidents and accidents, and they know that mental disease is the second most common cause of premature career temination among pilots.

The USA pilot association ALPA, recognised these problems some years ago and with the cooperation of airlines and medical authorities, devised a system of Pilot Advisory Groups (PAGs) to provide a counseling service. It found in many cases that pilots' health could be restored and their careers saved if they were afforded proper counseling at an early stage of their mental disability. Members of a PAG are carefully chosen and receive some training, but it is regarded as essential that they all be members of the peer group and that normal rules of industrial conduct be suspended if their advice is to be accepted by the pilot in need of advice. Case references may arise from the individual concerned, his family, colleagues or employer and many references are made when signs of alcoholism are detected.

Human factors in the selection and training of pilots

In addition to the applicant for pilot training having to meet the required medical standards for issue of a professional pilot license, attention is given to establishing the probability of the applicant continuing to meet the standards until reaching the generally agreed maximum age of 60 years. Efforts are made to establish the applicant's mental and psychological fitness to be a pilot, and his or her educational achievements and lifestyle are valuable in making that part of the assessment. Total abstinence from taking harmful drugs is required, and only a moderate use of alcohol is acceptable. An excessive use of tobacco is discouraged for it is known that the combined effects of smoking and cabin altitudes of four to ten thousand feet are harmful to visual acuity and other factors affecting general wellbeing, such as alertnesss and resistance to fatigue.

In mid-1996 it was announced that pilot selection tests used by the UK's RAF will be made available on a commercial basis to civil airlines and/or intending pilots with a view to avoiding costly training on persons unlikely to succeed in this career. A requirement of prime importance to an employer is compatibility with other crew members and possession of the special type of personality and discipline required to function in a supporting role as a copilot, but with the ability to assume responsibility to intervene if the captain appears to be making a critical mistake. These particular qualities will be looked for by the airline's training department rather than by the medical examiner, but are of great importance if the interface between flight deck crew and the airplane is to be optimum, for there have been a number of accidents caused by failures of crew coordination.

Pilots are taught to fly on simple types of airplane to permit them to acquire the ability to perform tasks requiring coordination of feet, hands and arms in order to maneuver the airplane in response to 'cues' obtained from the environment and aircraft instruments. A pupil pilot has to learn to simultaneously fly the aircraft and assimilate complex advice and information from the instructor – acting as both teacher and crew member – and from air traffic controllers, and to process that information before taking decisions and executing them in the best possible manner. It is during these early days that habits are formed, and the essential mental 'set' acquired that has the good pilot testing and questioning all information provided to him or her. Progression is then made from simple, single-pilot, single-engine types to more complex multi-crew, multi-engined types of airplane.

The aircraft systems provided to make these tasks possible are complex electro-mechanical or electronic systems powered from hydraulic, pneumatic or electrical sources and backed up by increasingly sophisticated display systems. The would-be airline pilot therefore has to assimilate and apply knowledge of a technical nature. Pilots' early training is when they first become aware of the need to adjust skills and mental attitudes in order to make the best use of the performance of the aircraft, and to be able to contend with any deficiencies it may have. A widespread use of computer-based training devices, flight trainers and flight simulators has enabled these learning tasks to be performed in safety and at an acceptable cost. Some experienced pilots believe that the inherent safety of the flight simulator is its only deficiency as a training aid, for it cannot produce in a pilot undergoing training, the stress that arises from the risk to life that is always present in flight training.

It is in the interests of airlines to select pilots who 'fit' the company culture, because it has been noted that when pilots of different backgrounds are put together, difficulties may arise even though the basic skills of pilotage are dictated by technology. At a 1995 IATA seminar on human factors, it was reported that one airline employed 250 pilots of 36 nationalities, with 60% being from Anglo countries. Difficulties arise when different cultures clash over such matters as the authority of a captain. In Anglo cultures, a copilot is more ready to advise the captain of an observed mistake than in other cultures where the captain is not corrected as 'It would shame both of us'. In 1996, IATA proposed a study of culture/safety relationships. Even in the same country when airlines merge, serious and unforeseen difficulties can arise. The most extreme example of airline awareness of this problem is in Southwest Airlines USA, where part of the selection procedure includes an interview with a group of airline employees.

Trends in airline pilot training

Doubts are expressed as to whether the best possible training is given to airline pilots by airlines, after considering that more than 60% of accidents are human-factor oriented as a result of improvements in aircraft and systems not being matched by a similar improvement in crew performance. There are many cases where the crew do not select the correct procedure for the given situation, or the procedure is incorrectly performed. Training, therefore, remains a fertile area where improvements can be made.

By the 1980s emphasis switched from training and checking individuals to a concept of training as a crew, for it was recognised that the introduction of automated systems, reduced crew complement and increased air traffic had brought about a need to reconsider and redistribute cockpit workload. It was probably this reconsideration that caused the FAA to require that a captain and copilot cannot be 'paired' unless at least one of them has a minimum of 75 hours of line operating experience on the type of airplane *and* that each captain and copilot must acquire 100 hours of line operating experience for consolidation of knowledge and skills, within 120 days of completing a proficiency check on the new airplane type.

In 1993 a French inquiry into a major fatal accident found that regulations governing airline pilot type-conversion training for 'latest generation' airliner flight decks were totally inadequate. It was perhaps a realisation of these differing standards of training and achieved levels of pilot competence and its investigation of a 1994 accident to a regional airliner, that were catalysts leading the USA's National Transportation Safety Board (NTSB) in late 1995 to call for disclosure of pilots' training and work records – a proposal that is sure to lead to complex legal issues being raised. USA ALPA pointed out that a failing pilot may reflect a failed training program, and that a more useful step would be to do a better job of screening new applicants who do not have an established training history or records!

Where airline pilots already have professional qualifications, maximum use is made of computer-based training, where the pilot uses a personal computer to learn data and systems at his or her own pace. This proceeds to cockpit procedures trainers, fixed-base simulators and finally to full flight simulators to learn operating procedures, flying skills and the skills needed to operate as a team. Only a minimum use is made of expensive in-flight training. Training experts recommend that attention be given to full flight simulator training in: windshear encounters, pilot incapacitation, cockpit resource management (CRM), emergency descents, rapid decompression, cockpit fire and smoke procedures, ditchings and abandoned take-offs for reasons of landing gear problems, flight control failures as well as engine failure.

In 1995, ICAO adopted an annex amendment that introduced mandatory human factors training for airline flight crews, a far-reaching step that obligates more than 180 states to take this action. Early human factors training was aimed at the individual, but modern (1996) programs address a systemic picture and educate pilots in relation to systemic flaws that might trigger safety breakdowns. Pilots, doctors and psychologists pool their respective expertise to develop a curriculum biased toward physiology, psychology and conventional man-machine interface problems. The assumption in this approach is that individuals cause accidents but a search must be made for failures in the social and organisational dimensions of the air transport system. By the 1990s crews were being taught the importance of making the best use of cockpit resources, by a technique known as cockpit resource management (CRM).

Crew coordination

Good coordination between flight crew members is essential if the highest possible standards of safety are to be achieved. A copilot must be trained to perform his or her own designated duties, to provide whatever assistance the captain requires and, more importantly, to act as a monitor of the captain's performance. For these reasons pilots should be trained to the same standards as the captain if their credibility in these different roles is to be accepted. Western airlines instruct copilots that they must bring *all* deviations from standard operating procedures, breaches of air traffic clearances or any event that might threaten safety, to the attention of the captain by means of prompt carefully phrased, respectful but firm questioning, and if that fails, to actively intervene to ensure safety. It is easier to propose such a procedure than to implement it in the air, for in almost every circumstance the captain will be the person with the greater experience and higher status in the airline hierachy. On those occasions when the copilot is flying the aircraft, the captain has only to state 'I have control' for the copilot to revert to his normal support role. So far as is known, differences of opinion between pilots have not yet degenerated to 'wrestling' to gain or retain control. However, investigation of an accident in 1993, when a DC10 left a runway on landing causing $35m of damage, revealed that when the copilot flying the airplane announced his intention to *go around* the captain stated 'No no no, I got it'. The copilot relinquished control but the captain landed the airplane well down the runway, eventually leaving the paved surface.

New systems of confidential and anonymous reporting of incidents have brought many instances of poor crew coordination to light, and airlines and pilots' associations strongly support the reporting systems and improved flight deck procedures that result from studies made of the reports.

Some reports give 'unacceptable flight deck behaviour' as the main problem, and examples given of such behaviour are: inefficiency, passivity, laziness, incompetence, boredom, not monitoring other persons' actions, and absences from the flight deck. For captains, difficult crew members are those who have strong and obstructive personalities, or who have only a passive involvement with their duties. For copilots, the difficult personality is the captain with an abrasive and intransigent personality and an autocratic style of leadership.

Major airlines have adopted 'Line Oriented Flight Training' (LOFT), where the flight deck crew 'fly' a complete trip in a high-fidelity flight simulator, so providing a check on performance as a crew in a realistically simulated environment. Good results are obtained by this method, and it is generally agreed to be better than the previous method of checking the performance of individuals on only segments of a simulated flight. A number of emergencies are programed into the 'flight', and consequences of any errors are carried forward into the flight as they would be in real life. A video recording playback facility enables the performance of crew and individuals to be analysed in joint post 'flight' discussion.

Cockpit resource management

A 1995 FAA circular defined cockpit resource management (CRM) as:

'one way of addressing the challenge of optimising the human/ machine interface and accompanying interpersonal activities. These activities include team building and maintenance, information transfer, problem solving, decision making, maintaining situational awareness and dealing with automated systems.'

Equally valid descriptions use terms such as:

■ Situational awareness
■ Leadership
■ Use of standard operating procedures and check lists
■ Delegation of tasks
■ Assignment of responsibilities
■ Establishment of priorities
■ Use of information
■ Communication
■ Monitoring and cross checking
■ Problem assessment
■ Distraction management

The Flight Safety Foundation has stated that:

'CRM is an embedded operational behaviour, it should be introduced at the earliest stage (ab initio) of a pilot's education and then integrated into the routine of training throughout the pilot's career'.

It is clear then that CRM is regarded as a major contributor to the fight to improve air safety.

'Fitness' to fly

Ordinary short term illnesses and disabilities affect all members of the community, but their effects on pilots are more serious and the pilot should not fly when common illnesses that are merely a nuisance in many occupations, may be adverse to the interests of air safety. The common cold frequently affects resistance to fatigue, and has an adverse effect on alertness and visual acuity. Stomach upsets, food poisoning and diarrhea are the most frequent causes of pilot in-flight incapacitation, and the inability of a pilot to perform his or her duties during the course of a flight is a serious matter considering the minimum crew complements that are now common. There have been instances of unqualified persons being called upon to assist the unaffected crew member(s) to complete a flight, and in the case of two pilot only crews, the remaining pilot in some cases has to use 'emergency' procedures in order to ensure safety.

Major airlines recognise this problem and one of the ways in which they reduce the likelihood of food poisoning hazarding a flight, is to arrange for flight crew members to eat different meals in flight. Airline medical departments provide advice to the crews on what medications may safely be used, and some cold cures which contain anti-histamines are not prescribed for use by pilots because of their potentially harmful side effects. Another serious problem is that of a pilot who has too great an intake of alcohol before a flight, and almost all airlines apply an 'eight hour rule' that prohibits use of any alcohol within eight hours of the commencement of a flight.

A number of studies have been made of the threat posed to air safety by pilot incapacitation. Generally it is accepted that the medical and operational aspects of the problem must be considered together, with medical standards and healthcare seeking to prevent pilot incapacitation to the maximum degree possible, and training seeking to minimise the operational and safety consequences of incapacitations that do occur. Three types of incapacitation are considered, obvious, subtle and cognitive, and all three types can occur in all age groups and may be temporary or permanent, partial or complete. Several studies and questionnaires have established the following rank order of reasons for pilot incapacitation:

(1)	Diarrhea	(9)	Severe back pain
(2)	Nausea	(10)	Toothache
(3)	Stomach cramps	(11)	Weakness or faintness
(4)	Rapid onset of any other pain	(12)	Leg or foot cramps
(5)	Abdominal pain	(13)	Severe nose bleed
(6)	Headache	(14)	Severe sneezing spell
(7)	Other	(15)	Severe coughing spell
(8)	Earache	(16)	Severe chest pains

It was studies of this type of problem that led airlines to simulate obvious and subtle incapacitations during flight simulator training sessions, and to devise cockpit procedures aimed at detecting subtle incapacitations. Most airlines now instruct crew members to regard any failure to respond adequately to *two* communications as being probably indicative of incapacitation. They are further taught that their response must be:

■ Maintain control of the airplane
■ Take care of the incapacitated crew member
■ Reorganise the cockpit and land the airplane

This training is successful and examination of the results of many simulator training sessions show that well trained crews cope effectively with the problem. Medical aspects of the problem are recognised in the United Kingdom by applying more strict medical standards to pilots who fly single crew aircraft than to pilots of multi-crew aircraft. The results of all this research, application of appropriate medical standards and improved pilot and crew training, has persuaded some authorities that the risk of a fatal accident arising as a result of pilot incapacitation is no greater than 1 : 100 million to 1 : 1000 million, which if true is highly satisfactory.

Statistical studies made by pilot associations and other organisations show that the risk of a pilot dying in flight is very small, and in a ten year period there were only 15 cockpit deaths in the USA causing one fatal accident. Internationally there were 17 cockpit deaths in a seven year period resulting in five fatal accidents with 148 casualties.

In a USA ALPA study, a questionnaire sent to 14 000 pilots showed 1500 cases of pilot incapacitation and 28% of pilots involved in these events believed that the safety of their flights were adversely affected by the incidents. Operationally there is little difference between a death and an incapacitation and the two problems are handled by pilot training.

Drugs and alcohol

There have been few detected instances of drug abuse by airline pilots, and a system of frequent random testing for drugs introduced in the USA in 1990 provoked angry responses from USA ALPA, that believes the practice to be an unjustified abuse of civil liberties. The extremely few cases detected by the checks (only two positive flight crew member tests out of 30 732 random tests conducted in the first six months – a rate of 0.007%) led to a reduction in the original 50% testing carried out, but ALPA has not yet succeeded in having the number of tests reduced to its suggested 10%. ALPA argued that 50% random testing was wasteful at a cost of $24 m per year, and that a 10% rate would be an ample deterrent. It further pointed out that when the FAA reduced testing of its own employees from 50% to 25%, it had no effect on the effectiveness of the program. ALPA does *not* oppose drug testing for applicants for aviation safety critical jobs – pilots, air traffic controllers, aircraft dispatchers, aircraft maintenance and ground security employees, etc.

Subsequent to introducing the random drug test procedure, the FAA introduced random tests for alcohol abuse in January 1995, following upon a well publicised case of three pilots being accused of flying a commercial jet while intoxicated. The rule prohibits: acting as a crew member within eight hours of using alcohol, while having an alcohol concentration of 0.04 or greater, using alcohol within eight hours of an accident or until tested, and refusing to submit to an alcohol test. The regulations require employers to permanently ban pilots from flying, 'who have been determined to have used alcohol or a controlled substance while on duty, twice violated the alcohol related conduct prohibitions or failed two required drug tests'. The regulations provide no procedural due process protections, and the tests are of 25% of covered employees on an annual basis with random selection by the employer. ALPA is concerned that failing an alcohol test carried out by the highway police authorities, will lead to pilots being suspended from work as a pilot until cleared by the Federal Air Surgeon, and it believes that a combined rate of testing for drugs and alcohol should be limited to 10%.

Accidents possibly involving psychiatric problems in pilots

Psychiatric illness is far less common among airline pilots than in the general population, because of the careful selection procedures applied to pilots and the regular medical examinations they undergo. There have been few reported instances of psychiatric problems in pilots causing accidents.

In 1982, a jet airliner crashed in Tokyo Bay while on a landing approach. The aircraft commander had a two year history of illness that originally manifested itself in difficulty in maintaining the required standards of competency during check flights, and in unusual social behaviour. After treatment by a psychiatrist he was returned to duty as a copilot, and after about one year was permitted to resume duty as captain. The probable cause of the accident was stated to be that the captain, at low altitude during the landing approach, pushed forward on the control column, closed the throttles and at a later stage selected reverse thrust on two of the engines. It was not possible for the other crew members to retrieve the situation and the airplane crashed into Tokyo Bay with a resulting heavy loss of life.

The accident inquiry revealed that crew members on earlier flights had noticed some abnormality in the captain, but none of them took it to be a serious mental disease. Japan Air Lines has since stated that the accident was not prevented, because information regarding the mental condition of the captain was not properly passed to the doctors and top management of the airline, so preventing a proper judgment being made.

In 1994, a preliminary report of a Moroccan airline accident causing the loss of 44 lives, stated that the accident was caused by the pilot's suicide, and claimed that this finding was substantiated by cockpit voice recorder transcript evidence. It was alleged that the captain disengaged the autopilot and pointed the aircraft at the ground with it breaking up before impact. This preliminary finding was disputed by two pilot associations. They stated that the suicide theory was based only on circumstantial evidence, and that the crew had reported trouble with the forward cargo door in the aircraft on the previous sector. They continued that only nine days before the accident, there had been a pilot strike because of disquiet about maintenance standards on the aircraft fleet.

These unhappy events may have caused an October 1994 meeting of the European Joint Aviation Authorities to call for the psychiatric testing of pilots. Opponents of the proposal countered by pointing out that the existing structure of line and recurrent training of pilots, and the availability of flight and cockpit voice recorders, already permit early detection of problems which result in under-performance, and that in any event there is no test available that is widely accepted.

Fatigue

In the early post World War II years, the design limitations and performance of airline airplanes provided a safeguard from crew fatigue because the aircraft could not be flown long distances without refuelling, and were not sufficiently well equipped to fly at night or in bad weather. Airplanes were scheduled to fly several short sectors each day on a long flight from Europe to Asia or USA to Africa, with aircraft, passengers and crew staying overnight at some suitable place and the journey taking several days to complete. The crew were thus able to maintain a near-normal pattern of activity and sleep, and to adjust to climatic and time-zone changes in a near ideal way.

The introduction of improved aircraft and weather radar enabled airlines to reduce their operating costs by scheduling the airplanes to fly from one end of a route to the other, with stops made only to refuel and change crews. At that stage of development of civil air transport, the frequency of flights on the busiest routes was very low, sometimes only two or three per week. The operations departments of airlines saw it as being their duty to schedule crews to fly as long as possible each 'day' in order to obtain an acceptable (to them) annual rate of utilisation. Crew scheduling departments took a simplistic view that the crews' daily work rate was the annual flying hours divided by 365, ignoring the fact that fatigue is cumulative but rest is not.

In 1980, in response to a Congressional request, NASA created a Fatigue/Jet Lag Program to determine the magnitude of the problem and its operational implications. Three program goals were established:

- To determine the extent of fatigue, sleep loss and circadian disruption in flight operations
- To determine the impact of fatigue on flight crew performance
- To develop and evaluate counter measures to mitigate the adverse effects of fatigue and maximise flight crew performance and alertness

Over a period of 12 years studies were conducted in a variety of aviation environments, controlled laboratory conditions and one full mission flight simulation study.

In one study, 74 pilots from two different airlines were examined before, during and after three and four day commercial short haul trips on the east coast of the USA. The pilots averaged 41.3 years of age, and 14.6 years of airline experience. They were subjected to measurements of core body temperature, heart rate, and motor activity rate every two minutes, and provided subjective ratings of fatigue and mood and recorded their sleep episodes and other activities: meals, exercise, duty time, etc. The next study was of 29 male flight crew members, average age 52 years, flying B747 aircraft on one of four commercial international route trip patterns with sectors over eight hours long. It was this study that provided evidence of the benefits of short 'naps' (*see Circadian rhythms* below), and both studies were precursors to the work of an Aviation Rule-making Advisory Committee (ARAC) that developed a draft advisory circular which was forwarded to the FAA.

Circadian rhythms

Aviation activities require a high state of alertness for safe crew performance, and this state of alertness varies with time of day in accordance with individual personal tendencies toward being a 'night' person or a 'day' person. Studies performed on shift workers reveal a marked increase in errors after midnight hours as compared to daytime hours. Evidence shows that the probability of making mistakes significantly increases during those times when the individual is accustomed (adapted) to sleeping but for various reasons has to undertake certain duties. Analyses of confidential reports to NASA's ASRS indicate that 21% of all reported incidents are related to fatigue, and tend to occur more frequently in the early hours of the morning, and are often potentially serious.

When undertaking trans-meridianal flight toward the east or west, travellers find their circadian rhythms progressively placed out of phase by the number of meridians crossed. The effect of this displacement is most significant when five or more time zones are crossed. Crew members find the long distance east-west destination time out-of-phase with their eating and sleeping desires and on their subsequent departure, may find themselves attempting to accomplish complex pre-flight planning at a time when physiology is calling for sleep.

In aviation, challenging and emergency situations can only be met by the closest margin of performance adequacy, and the circadian rhythm low point may compromise this margin to the point of human failure. It is therefore not surprising that prevention of crew fatigue continues to occupy the minds of those responsible for air safety.

After conducting a research program, NASA stated in 1990 that allowing pilots to take a rest in the cockpit during long haul flights 'may provide substantial safety benefits'. The project was conducted with the FAA and involved three man crews of United Air Lines and Northwest Airlines on scheduled trans-Pacific flights. Measurements were made of vigilance, sustained attention, activity levels, brain activity as recorded on an electro-encephalogram and subjective reports of mood. Researchers found that the crews who were allowed to take rest periods were generally able to fall asleep easily and sleep well and remained more vigilant particularly in the crucial final phase of the flight. Airlines in several countries have authorised this practice with the approval of their regulatory authorities.

Flight and duty time limitations

The importance of ensuring that flight crews do not suffer from fatigue to the extent that safety is jeopardised has been recognised by ICAO adopting as a Standard, a requirement that member states of that organisation adopt regulations aimed at preventing undue fatigue. It is regrettable that no guidance is provided on what limits the recommended regulations should prescribe, and the omission results in each state taking whatever action it chooses in this important matter. Regulations vary greatly from state to state, and in some instances there are virtually no constraints put upon the scheduling departments of airlines.

Among most reported items in incident reporting systems, are allegations by flight crews that they suffer fatigue to a degree that threatens air safety. In a number of cases, reports were made that *all* flight crew members fell asleep at the same time. Many reports state that the greatest difficulty arises not from any single duty period, but from work patterns that cause serious sleep deprivation by requiring that the normal rhythm of wakefulness/sleep is disrupted.

Circumstances causing such disruption arise in short-haul flying when alternate day/night working is scheduled, and in long-haul flying when the time interval between successive duty periods causes similar disruption, aggravated by changing local time on east-west flights. Flight crews are generally agreed that the worst work patterns in long-haul flying arise on worldwide east-west routes. In these cases successive aircraft leave their home base at 24 hour intervals and crews are scheduled to work for 12 hours and then to rest for 24 hours before the next aircraft in the sequence arrives. Such a rest period ensures that the crew normally sleep immediately after arrival when they are tired from the flight, and are then awake for a long period before taking their next flight. The sleeping period may coincide with daylight hours, resulting in exposure to wake stimulii, aggravated by the noise of normal life in the hotel being used. A graphical plot, of sleep versus work in such a pattern, would show that the biological, physiological and psychological clocks of the flight crews are greatly disturbed from the ideal 8 hour sleep/16 hour wakefulness routine of normal life. Pilot fatigue is frequently a causal factor in aircraft accidents, and some countries require the previous duty/rest cycles of the flight crew to be examined in accident investigations and then to be published in the official report.

The urgent need for internationally agreed regulations controlling flight and duty time was demonstrated in early 1996, when a British report into a fatal accident to an Air Algerie B737 gave pilot fatigue as being a major causal factor. It was decided that the UK would increase inspections of foreign-registered aircraft where there is any doubt that international safety standards are being met, thus following the lead of the USA's FAA.

In 1995, a distinguished group of NASA researchers submitted to the FAA its suggested 'Principles and Guidelines for Duty and Rest Scheduling in Commercial Aviation', and these were generally welcomed by national and international airline pilot organisations.

The document addressed:

- Duty
- Duty period
- Flight duty period
- Window of circadian low
- Cumulative effects
- Off-duty activities including commuting before a duty period

- Sleep loss
- Time of day/circadian physiology
- Continuous hours of wakefulness
- Physiological capabilities
- Extended flight duty periods and compensatory off-duty periods
- Off-duty periods
- Recovery
- Definitions

A second document provided specific scientific references supporting the principles and guidelines in the first. The NASA scientists believe that two week, 30 day and yearly cumulative limits should be applied and that crew members should have the opportunity to sleep prior to an assigned duty period, though this may cause difficulties when pilots are on reserve (standby) duty.

The NASA experts' proposals were significantly modified by the FAA in its formulation of a Notice of Proposed Rule Making (NPRM), and the consultative process was still in progress in 1996. By March 1996, many regional airlines and general aviation operators were pleading that their particular circumstances required that they be exempt from the most onerous of the proposals, and the responses of major airlines were similarly influenced by increased cost of the proposed regulations. USA ALPA welcomed the proposed introduction of reduced duty hours for two pilot crews, increased rest periods, limits to extension of duty days to permit operational delays, restrictions placed on the duration of periods on reserve (standby) duties and the inclusion of ferry and positioning flights. They deplored, however, the absence of accountability for circadian physiology.

An extended discussion period, therefore, is resulting in increased lobbying activity by those for or against the NPRM, and the FAA is under severe pressure from all sides of the industry. This shows that the problem of regulating flight and duty time will not go away, and public interest in aviation safety and the pressures exerted by pilot associations and others will keep the matter active until a generally accepted solution is found.

In Europe, attention focused on the 24 member state Joint Aviation Authorities (JAAs) which have draft flight and duty time proposals that are suspect, because they had little input from aviation medical experts. They also seem to be *politically* motivated in that they are an attempt to find an agreement that is acceptable to all the 24 states aviation regulatory bodies. This has resulted in exclusion of the more restrictive regulations of some major states, in order to achieve acceptance by those minor states that have traditionally placed few restrictions on their air-lines. In all instances, the JAA proposals would result in *increased* flight and duty time (in some cases of up to three hours) for pilots covered by the present United Kingdom, German, Dutch, French and other reg-ulations. Because the JAAs have no legal status and pilot organisations are opposing the proposals, the fight may go to the European Community in Brussels and even to the European Court of Justice.

In 1995, according to the Canadian Air Line Pilot Association, Transport Canada published proposals that would impose annual maximum working hours 20% longer than those proposed by the JAA, and weekly maxima nearly double. It is probable, however, that Canada will be influenced by whatever the FAA decides, because US airlines are unlikely to accept that Canadian airlines should have a significant cost benefit at a time when the North American Free Trade Association (NAFTA), is attempting to remove trade barriers.

Ageing and human performance

Everyone has elderly acquaintances who remain remarkably young in their lifestyle, habits and standards of performance in tasks that may be demanding in terms of mental and physical effort. Everyone also knows people who are much younger, but who cannot perform these tasks to the same standard. The human race, therefore, does not have a single rate of ageing, and the regulatory authorities in civil aviation are not agreed on a single method of determining how long a pilot may continue to exercise the privileges of a license.

An ICAO Standard states that a professional pilot should give up a license to fly international routes at the age of 60. Some member states allow a pilot to continue beyond that age, and some airlines set pilot retirement limits at ages lower than state licensing limits for socio-industrial purposes. Safeguards provided to maintain air safety are the medical and competency checks professional pilots undergo at fixed time intervals. Six-monthly competency checks are carried out to standards set by the state, and pilot medical standards are internationally agreed. After reaching the age of 40 years, most pilots undergo two proficiency checks and two medical checks per year, with the medical checks including X-ray, ECG and EEG examinations.

When in 1959 the USA's FAA set 60 as the maximum age for professional pilots, it had concluded that the ageing process results in various deteriorations. These include a loss of ability to perform highly skilled tasks at speed, to resist fatigue, to unlearn and discard old techniques and to apply the rapid judgment needed in changing and emergency situations.

In 1995, the United Kingdom joined those ICAO member states that allow licensed pilots to continue public transport operations to the age of 65, and published a list of 60 countries that accept this declared 'difference' from the ICAO Standard. Nineteen other countries including the USA have refused to give 'blanket' authority for these pilots over the age of 60 to operate in their airspace, and a further 14 countries had not made a decision by the end of 1995.

Scientific tests have shown that cerebral functions slow down with increasing age, with the time required to press a button, after an optical or acoustic signal is provided, increasing by about 20% between the ages of 20 and 60 years. For a pilot it is necessary to coordinate cerebral activity and visual function, and tests have shown that the minimum time taken to obtain information from an instrument is at least 100 milliseconds. With increasing age, less information per unit of time may be physiologically stored and recalled, and the capacity of learning is reduced.

IFALPA informed ICAO that it believes an upper age for pilots should not be specified because safeguards to prevent pilots' age being a factor in air safety are already in place. These include: biannual medical and proficiency checks, training and checking programs designed to detect anything that may present a hazard, incapacitation training and upper torso restraint to prevent an incapacitated pilot from obstructing the other pilot from assuming physical control of the airplane.

Using FAA data, including Civil Airmen Statistics for 1994, it can be shown that older pilots with ATP certificates had excellent safety records over the years 1983–1995, and that a case for early withdrawal of pilot licenses cannot be established. It is probable that any impairment of mental and physical abilities are more than compensated for by increased experience and more mature judgement, (see Table 2.1).

A NASA source shows accident rates of Class 1 and 2 pilots (see Table 2.2) with more than 1000 hours total time and 50 hours recent time.

These statistics imply that the safety record of airlines will worsen during periods of rapid expansion when median levels of pilot experience are reduced, and will improve during periods of relative stability.

Human factors in aircraft design

The most modern types of aircraft introduced into service show an interesting diversity of opinion between major manufacturers. Although

Table 2.1 Pilots in command with ATP certificates.

Age	Active pilots (1994)	Total no. of accidents	Accidents per 1000 pilots
		← 1983–95 →	
20–24	309	116	375.4045
25–29	5 486	353	6.4345
30–34	15 526	414	2.6665
35–39	19 987	507	2.5366
40–44	18 310	579	3.1622
45–49	20 510	545	2.6572
50–54	16 012	451	2.8166
55–59	11 500	288	2.5021

(N.B.: The above table assumes that the age distribution of pilots in 1994 was typical of the years 1983–95).

agreed on a basic concept of providing the crews with information needed to carry out their tasks, manufacturers are not agreed on where boundaries are to be drawn between automatic or human control of the aircraft and their increasingly complex systems. The jargon phrase used in the argument concerning what information to provide to pilots is 'the need to know'. This presumes that the designer can foresee all possible circumstances and malfunctions, and prearrange the way in which the automatic systems or the crew will handle the problem, and so decide what to display and what to 'hide' from the pilot. The philosophy adopted by Boeing, is to keep the pilot 'in the loop' by providing full and unambiguous information and allowing him or her to take any required action. The Airbus philosophy has been to filter and reduce the amount of information provided to pilots, and in some cases prevent them from intervening in and/or over-riding the automatic systems.

Should designers decide that automatic systems are to manage the malfunction, they may withhold almost all information on the circumstances from the crew, and this philosophy is a source of controversy between manufacturers, regulatory authorities, and pilot associations.

Table 2.2 USA pilot accident rates related to age.

Age group	Accidents per 100 000 pilot flight hours
20–29	5.0
30–39	4.0
40–49	2.6
50–59	2.4
60–69	4.9

New and major problems are the certification of the software that will perform critically important safety-related tasks in automatic systems, and how to relate the hardware, software and 'human-ware' involved in the conduct of flight operations. It is already evident that the task of testing aircraft structures, systems and engines is much more simple than the testing of software for computer systems.

The training of airline pilots is greatly affected by decisions taken in this particular field, for a major part of pilot training has always been devoted to the successful management of emergencies and failures. Recent accidents have shown that there is an increasing tendency for crews to place undue reliance on the performance of automatic systems. In one case a crew persisted in the use of the auto-throttle system although the speed of the aircraft was too fast to permit a safe landing to be made, causing the aircraft to run off the end of the runway and into the sea. In other cases, crews have failed to check the performance of automatic navigation systems and have allowed aircraft to stray many nautical miles off course, resulting in exposure to the risk of collision or interception by military aircraft. It is apparent that the interfaces between pilot and automatic machines, and pilot workload need urgent study by qualified human factors specialists resulting in improvements in aircraft and systems design.

Airline pilots like modern flight instruments, systems displays and autopilots, but flight management systems (FMS) are much more controversial because of the excessive head-down workload required to reprogram them when a flight plan is changed, particularly in the terminal area. The most damning criticism is that the task is like working with DOS when Windows should be available!

3: Machines and Air Safety

From the very beginnings of aviation, problems involving the design, construction, repair, maintenance and operation of aircraft to as high a standard as practicable for the purpose of achieving safe and economic flight, have taxed the abilities of man. Much has been learned from early failures, and aviation museums contain many descriptions, drawings and some photographs, of early aircraft that failed to meet the objectives of their designers. Some failures were the result of bad design, some the result of a poor choice of materials and others of failure to understand the environment in which the aircraft was to operate.

Airworthiness of airline aircraft

The most severe conditions encountered in the earth's atmosphere, or the stresses produced by loss of control while in flight, are capable of destroying any aircraft structure that has yet been built, and these cases are considered in the design process. For example, a 'design gust' has been agreed internationally and aircraft structures are designed, built and demonstrated to be sufficiently strong to withstand the loads imposed by encounters with such gusts.

Airworthiness authorities accept that real gusts may infrequently be encountered in flight that exceed the 'design gust' in intensity. Theories of statistical probability are used to 'guarantee' that an aircraft structure will not suffer a critical failure more frequently than would cause the destruction of an aircraft more than once per 10 000 000 flights, from *all* causes! An additional and complicating factor of recent recognition, is the knowledge that an ageing process applies to aircraft structures as it also does to the human body.

The 'design gust' is one of 66 feet per second (fps), and it is assumed that the aircraft suddenly encounters a column of air that is moving vertically, and it is also assumed that the aircraft has been slowed to a speed known as V_b, (V being speed and $_b$ being buffet), before the encounter. If these assumptions are met, the strength of the airframe is sufficient to withstand the gust. However, should the aircraft be flying at cruise speed (V_c) at the time of the encounter, the 'design gust' is only 50 fps, and if the aircraft is at dive speed (V_d), the 'design gust' is only 25 fps. This 'tailoring' of the 'design gust' creates a requirement for the forecasting and airborne detection of vertical gusts. They are shown on meteorological charts as vertical wind shear, and are plotted in values of knots per thousands of feet. Encounters with a 'design gust' are extremely infrequent, and pilots who accumulate as much as 20 000 hours of flight time may never experience the phenomena (see Figure 3.1).

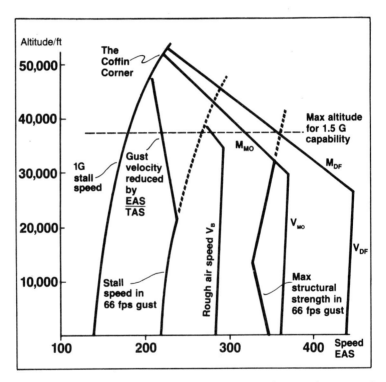

Figure 3.1 Limiting speed envelope showing stall speed, gust velocity, rough air speed, maximum speeds and altitude (*Courtesy of Captain T. Foxworth*).

Design criteria to achieve an agreed level of structural integrity for the purpose of achieving safety are well understood, and have been agreed on an international basis. Specimen aircraft structures are tested before manufacture starts in order to demonstrate that agreed safety factors, or margins, are provided. However, the demands of competition and the need to produce aircraft that are economic to operate require that the structure be as light as possible, and it is therefore only as strong as the agreed requirements. It frequently happens that the first test specimen fails, for designers know that they can always strengthen a structure if necessary. Early failure of a test specimen is not therefore regarded as a failure of design; on the contrary, a specimen that greatly exceeds the requirements will be regarded as being too heavy for the design task and likely to prejudice the economic viability of the aircraft.

Many factors are taken into consideration during the design process, but a basic principle that has achieved wide acceptance is to design structures that fail-safe, rather than have a safe life as was formerly the case. Fail-safe requires that no single failure may hazard the aircraft, and that structures provide multi-load paths to provide a redundancy feature whereby a second frame, spar, rib, or stringer can accept the loads carried by a failed part.

It is essential that transport aircraft continue to operate safely for many thousands of flights and/or many thousands of hours. It is normal for aircraft used on long flight sectors to achieve more than 4000 flight hours per year, and to be operated for as long as 20 years; figures unthought of 30 years ago. The servicing/maintenance process is therefore vital if these extended aircraft lives are to be safely achieved, and incipient failures identified and rectified before safety is prejudiced. Design for 'maintainability' is of great importance.

In addition to accepting loads imposed by gusts and flight maneuvers, structures have to be able to withstand bird-strikes, encounters with hail, lightning strikes and the onset of corrosion and fatigue. All these factors have caused aircraft to be destroyed on rare occasions. It is accepted, therefore, that a high standard of airworthiness can be achieved only by the dedicated efforts of many professionals. This starts with the adoption of high standards of design requirements by the airworthiness authorities, good design, thorough testing of specimen parts and the whole aircraft, careful manufacture with good standards of quality control, continuing high standards of maintenance and servicing. It continues with the exchange of technical information between operators and manufacturers, and the adoption of good operating practices by the user airline(s) and their pilots. This should continue throughout the life of the aircraft and particularly after a change of role or significant modification or repair of the aircraft, or transfer to another operator.

Regulatory controls

ICAO sets minimum requirements for the airworthiness and continuing airworthiness of aircraft, and detailed codes are devised and applied by states that manufacture, repair or operate aircraft. The FAA's Federal Aviation Regulations (FARs) are adopted by many states as their code, and airplane manufacturers would welcome the adoption of a single system to be applied internationally. The major 'competitors' to the FAA are the European Joint Aviation Authorities (JAAs), comprising 24 states including a number active in manufacturing aircraft. The JAAs are not a legally constituted body, but act with the concurrence of member national authorities, although each reserves the right to impose additional requirements. The JAAs' airworthiness codes were developed from FARs for airframes and British requirements for engines. Under strong pressure from manufacturers complaining of the need to build to two codes, or modify aircraft sold to particular states, the FAA and JAA are making efforts to remove differences between their codes. In 1992 the USA General Accounting Office reported to a US Congress sub-committee that slow progress was costing manufacturers 'millions of dollars' and pointed out that of 267 'differences' existing ten years earlier, 122 still existed. Since that date better progress has been made.

A major difficulty delaying this process has been an FAA practice of allowing manufacturers to use 'grandfather rights'. An example causing great difficulties was when the FAA authorised the B747 – 400 to be designed and built to requirements applied to the original B747 some twenty years earlier. This was notwithstanding findings of accident inquiries, operators' experience suggesting that improved requirements should apply and the FAA's own advances in safety regulation. The rationale used to justify this practice is that the later model is a 'derivative' of the original model and *not* a new type, and that the original model achieved an acceptable record of safety.

The JAAs insisted that the B747 – 400 meet a new requirement for increased strength of the cockpit floor to protect flight control runs from the effect of a decompression. All airplanes sold to European operators are built to this requirement, held to be necessary because of accidents to airplanes of several types. Another 'grandfather rights' difficulty arose when it was suggested that late model 'derivative' B737s be certificated to carry more passengers than current regulations would permit, while new types being built by Airbus were restricted to fewer passengers although the passenger cabin size and emergency exits were near identical.

Since 1993 there has been a JAA/FAA interim policy on 'the joint type-certification basis for derivative large airplanes', which describes amendments to Federal Aviation Regulations/Joint Air Regulations (FARs/JARs) allowing regulators to implement some of the proposed new policies and putting pressure on manufacturers to be ready for them. It is believed that although minor irritations may continue, good progress is being made to harmonise requirements, although European regulators claim to be concerned that the FAA's twin roles of regulating safety issues *and* promoting the economic health of US industry may lead to further conflicts.

Less confidence is expressed about achieving airworthiness agreements with Russia, and both the FAA and JAAs have experienced difficulty because the design philosophy, standards, procedures and oversight practices in the Commonwealth of Independent States (CIS) are very different from those with which the West is familiar.

A peculiarity of FARs has been differences existing between Part 121 applicable to large scheduled airliners, and Part 135 applicable to commuter and general aviation airplanes. These differences have affected both airworthiness and operational matters, with Part 135 regulations being markedly inferior to Part 121. As a result of a number of fatal accidents to scheduled commuter carriers, the FAA in 1996 placed these operations *with more than ten seats* under Part 121, making the operations much more safe and satisfying concerns voiced by the public and by ALPA. The newly applied regulations affect: airplane systems and emergency equipment, aircraft performance, crew training, flight and duty times, manuals, carry-on luggage, de-icing, despatch, provision of weather radar and maintenance standards. These improved standards are welcome, but it is surprising that cargo airplanes as large as B747s remain exempt from some Part 121 requirements such as carrying the Traffic–alert and Collision Avoidance System (TCAS).

A new and welcome initiative promising benefit to air safety is that the FAA is taking action against the airlines of countries judged not to comply with international safety-oversight standards. By the end of 1995, of the 50 countries evaluated under the International Aviation Safety Assessment (IASA) program, 24 had been judged deficient, with bilateral air service agreements with 11 Category 3 countries being terminated (airline operations to the USA being suspended), and bilaterals with a further 13 Category 2 countries being 'frozen' while they work to correct their deficiencies. A Category 2 rating means that new aircraft and routes cannot be added to operations to the USA. A number of these countries are in Latin America, some in the Caribbean and fewer from other regions. The IASA takes operational matters into account in addition to airworthiness and, although viewed as punitive by the affected countries and airlines, it should be welcomed by air passengers. It should achieve a drastic diminution of gross differences between achieved safety standards where – in the 1970s/80s – the risk of being involved in a serious accident was 40 times higher using South American airlines compared to those of the USA.

ICAO responded to the initiative taken by the FAA by stating that its technical cooperation programs are now aimed at assisting member states to reach the organisation's standards, and the USA, welcoming this change, offered to lend experienced personnel to assist ICAO. The European JAAs have stated that they will adopt similar measures to those of the FAA, and the United Kingdom began, in 1996, to increase surveillance of foreign registered air operations (charter and wet lease, flight crew included) following a fatal accident to an Algerian registered airliner in the UK. In 1996 Germany was proposing an air safety blacklist of airlines that have questionable safety standards, and banning them from Europe. This proposal followed the crash of a Turkish registered B757 off Dominica in the Caribbean when 189 persons were killed; mainly German holidaymakers.

Ageing aircraft

In May 1977 the port tailplane of a Boeing 707 failed during an approach to land, and all the six crew of the all-cargo aircraft was killed. The tailplane had been designed on fail-safe principles and a catastrophic failure was, in theory, impossible as the fatigue crack should have been detected before it became critical (see Figure 3.2). The aircraft had flown 47 600 hours while making 16 700 flights, and the accident is regarded as one of the first caused by the age of an aircraft.

It was a fatal accident to a B737 in Hawaii in April 1988 caused by a major part of the cabin roof structure separating from the airplane, that was the catalyst that saw airworthiness authorities, aircraft manufacturers and airlines get together to devise safety audits and remedial work programs for ageing jets. Mandatory inspection and structural work programs were introduced for all types of aircraft approaching the end of their design lives, or having a history of structural defects (cracks/corrosion), and these programs continue. Table 3.1 shows that many airplanes have exceeded design limit hours and/or cycles, but safely continue their lives. Many will be retired when noise limits ground them, or cost of maintenance becomes excessive or when the improved economics of new designs become compelling to operators.

Boeing has a policy of buying old airplanes from operators to tear them apart, or alternatively, to test them beyond their design lives to determine where their new airplanes may be improved. It also arranges

Figure 3.2 B707 tailplane failure (*Courtesy of UK Accidents Investigation Board*).

to be present when 'high time' airplanes undergo normal heavy maintenance, and has by this means surveyed 185 airplanes from 95 operators. This has resulted in improved anti-corrosion features being
introduced to new airplanes that are designed to have longer lives. As an
example the B777 has a minimum design service objective of 44 000
cycles and will be fatigue-tested to twice that figure.

In late 1995 there were 5671 jet powered airplanes and 2063 turboprops in service that were more than fifteen years old, some of which may
be temporarily parked. Most of the 400 parked jet powered airplanes are
unlikely to return to service because they cannot meet Stage 1 and Stage
2 noise requirements, but 39% of parked airplanes are Stage 3 compliant
and presumably may be returned to service. Table 3.1 shows 20-year old
jet powered airplanes in sevice in 1995.

A problem that has recently arisen is of civil aircraft approaching the
end of their safe lives, being retired by major airlines and then being

Table 3.1 Twenty-year old jet airplanes.

Type	No.	Design life (hours)	Fleet leader	Design life (cycles)	Fleet leader
Concorde	4	45 000	18 778	24 000	6 766
B747	205	60 000	97 424	20 000	32 517
DC10	179	60 000	91 424	42 000	36 207
Tristar	97	115 000	72 202	210 000	33 175
A300	16	60 000	48 162	36 000	35 001
DC8	276	100 000	86 760	50 000	48 146
B707	146	60 000	89 295	20 000	42 169
IL62	28	30 000	39 026	12 000	8 225
B727	757	50 000	78 487	60 000	78 490
Tu154	79	30 000	38 131	15 000	16 990
BAC111	81	85 000	62 050	85 000	80 672
DC9	637	75 000	79 062	100 000	99 410
B737	325	51 000	80 638	75 000	90 044
F28	59	60 000	64 644	90 000	89 652

bought by developing world airlines. Some of these airplanes need major maintenance work performed, but the new owners may not have the technical and financial resources to keep the aircraft airworthy. The problem is increased by a small but growing tendency for carriers that cannot or will not meet the requirements of states that enforce safety regulations, moving their operations to other states that are less demanding. This is a case of air transport following the bad example of sea transport, and adopting the 'flag of convenience' technique to reduce costs by evading safety regulations.

Bird strikes

A difficult problem faced by designers of aircraft and engines is to make their products capable of withstanding impact with birds in flight, and it is a design requirement to consider 'standard' birds to be encountered at the aircraft's design climb speed up to a stated altitude.

The risk incurred from in-flight encounters with birds was recognised as early as 3 April 1912, when C.P. Rodgers, the first person to pilot an aircraft across the US (taking 84 days, 12 crashes and scores of forced landings to complete the feat), took-off from Long Beach, California, ran into a flock of gulls and plunged to his death. He was the 147th person to die in an aircraft accident and the first to die from a bird strike.

Significantly increased costs and additional weight are incurred in the design and manufacture stages of large passenger aircraft, where items such as windshields, leading edges of wings, tailplanes and fins are all designed to survive the strikes from 'standard' birds in standard conditions without catastrophe. As an eight pound bird colliding with an airplane travelling at 300 miles per hour produces forces calculated at 40 000 pounds, it will be seen that designers face formidable tasks. A method used to verify aircraft and engine strength is to use a specially designed cannon to fire the carcass of a chicken at test specimens. It will never be possible to design aircraft to survive the worst possible cases of bird strikes, for hawks tracked by radar have been found to fly at 20 000 feet, and migrating waterfowl have been counted in flocks of 50 000. No aircraft could survive collisions with such targets while flying at cruising speed.

During the period 1980 through to 1995, ICAO received reports of 62 000 bird strikes from 177 member states, of which more than 50% occurred during approach and landing and 40% occurred during take-off and initial climb. Of these bird strikes, 2030 had an effect on flights (abandoned take-offs or precautionary landings, and 3285 caused engine damage or shut-downs. Typical bird strike incidents include a B727 taking off from Winnipeg, striking six Canada geese at 1500 feet above ground level, losing an engine, suffering severe damage to the fuselage and wing, with one of the birds penetrating the leading edge of the horizontal stabiliser (see Figure 3.4). In another incident a B747 taking off in fog ingested a gull into an engine and suffered $2 million damage (see Figure 3.3).

In the USA in 1994, 2150 bird strikes were reported causing 112 precautionary landings, 66 aborted take-offs and damage to 118 engines. JFK International Airport at New York suffers from colonies of laughing gulls, with the bird population increasing from 15 nesting pairs in 1979 to 7629 pairs in 1990. This caused 100 to 315 airplanes to be struck by 110 to 391 birds per year during the period from 1979 to 1992. In 1991 and 1992 two B747s at JFK suffered strikes during take-off, one aircraft abandoned take-off and required the replacement of brakes and ten tires, and the other jettisoned 200 000 pounds of fuel and returned to JFK. In addition to programs designed to scare the birds away, a team was assigned to shoot gulls in the summers of 1991 through to 1993, with more than 35 000 gulls shot and a 70% reduction in bird strikes at the airport. Shooting is now greatly reduced as the birds have learned to stay outside the airport boundary and the breeding population is almost static.

Long grass on airports discourages birds, as does the use of recorded bird distress calls and pyrotechnics, but gas cannons, scarecrows, effigies of dead birds and falcons all have a limited use as birds quickly learn to

Figure 3.3 Wardair Boeing 747, Victoria, Canada. Gull bird strike – US$2 million damage (*Courtesy of Transport Canada*).

recognise false threats. Airports throughout the world suffer similar problems, putting lives at risk and causing the airlines millions of dollars in airplane/engine repairs. Remedial measures include the banning of garbage dumps and bird-supporting vegetation near airports.

The International Bird Strike Committee is based in Europe and its documentation shows the dimensions of the problem: an averaged rate of birdstrikes is 5 per 10 000 movements.

- A turbo-prop Lockheed Electra was totally destroyed at Boston in October 1960, being the worst bird strike accident with 62 lives lost
- In 1975, a DC10 on take-off was a total loss after collision with a flock of gulls at Kennedy, when No.3 engine disintegrated, fan blades ruptured a wing fuel tank and the airplane caught fire. All passengers on board were airline employees and escaped safely and very quickly
- In 1988, a B737 taking off in Ethiopia flew through a flock of pigeons and lost both engines. 35 lives were lost in a crash landing
- A BAe146 on a night-time take-off in Italy, collided with a flock of gulls, causing one engine to be shut down and the other three engines to surge and lose power. After landing, all four engines had to be changed and the airplane cleaned of its coating of blood and feathers
- Less serious, but almost as frightening, was when a lady passenger in

Figure 3.4 Air Canada B727, Winnipeg, Canada. Canada geese bird strike – US$300 000 damage (*Courtesy of Transport Canada*).

a forward first-class seat in a B747 was suddenly covered with blood when the plane hit a bird at 4000 feet after taking off from London Airport at night. The bird's remains were forced through the passenger's window. It is reported that she had a baby in her lap and thought that it had exploded!

■ The highest reported bird strike was at 37 000 feet over West Africa – a vulture

■ A B727 had a windshield damaged at 20 000 feet over Scotland

■ A JT8D engine on a B737 was struck by gulls. All the fan blades detached, the nose cowling fell off, and some debris fell to the runway causing minor damage to the other engine and to one cockpit window. Two engine-mounting bolts were broken and the engine was supported by only one bolt and two hydraulic pipes

■ A total of 190 deaths, 30 fatal accidents and the destruction of 52 civil aircraft are assigned to bird strikes.

New engines are required to be able to ingest mixtures of up to six 1.5lb birds, one 2.5lb bird and at least one bird weighing up to 8lbs depending on inlet areas. After ingestion, the engine must continue to operate for two minutes with no power lever movement, three minutes at 75% of take-off thrust and six minutes at 75% of maximum continuous thrust. The engine must contain any fragmented or fully separated blades.

Ditching

Emergency alighting on water, 'ditching', is rare but not unknown and it is surprising that large airplanes are no longer tested for ditching capability using scale models. The last successful ditching of a jet airliner was a DC9 in 1970 and of the 63 persons on board, 22 passengers and one cabin attendant died. In 1995 a RAF military derivative of the Comet airliner was successfully ditched in Scotland. The FAA and the UK Civil Aviation Authority accept proof of ditching ability by analysis or extrapolation from earlier tests. However, there are specialist organisations capable of conducting tests of models that might reveal poor ditching characteristics that could be improved by minor design changes.

Wheels, brakes and tires

These vitally important components suffer the worst possible treatment. They are as small and light in weight as possible and are subject to high temperatures during the taxying, take-off and landing phases of the flight. They are twisted and tormented during ground maneuvers, suffer rapid changes of temperature when, for example, they are cold-soaked for hours of cruising flight, and are expected to perform at maximum efficiency immediately the airplane lands. The pilot cannot normally see them, and usually has no information on their condition after the airplane begins to move under its own power. On the largest civil airliner, the B747-400, there are 16 main wheels and two nose wheels to support 873 000 pounds at ground speeds of up to 200 mph. There is therefore great potential for disaster in the event of failures.

Wheels have exploded in flight or on the ground, causing significant damage to aircraft structures and to anything else within range, in much the same way as shrapnel from bombs and shells. As early as 1973 Boeing was recommending that all braked wheels have tires inflated with nitrogen, in order to prevent spontaneous combustion of isoprene gas released from overheated tire liners as had occurred on a number of occasions. In June 1987, the FAA issued an Airworthiness Directive limiting the amount of oxygen in inflated tires to 5% by volume for most jet powered airline airplanes.

The critical role wheels and tires play in the safety of flight operations is shown by the fact that more than 24% of all rejected take-offs are due to tire failures (almost equalling the 26% due to engine failures) and that those rejected take-offs are four times more likely to result in an accident or incident than rejected take-offs for engine failure. Tire failure and the resultant in-flight fire on a Nationair DC8-61 at Jeddah in July 1981, resulted in the loss of 261 lives. In 1986 a seized brake caused a tire to explode in a Mexicana B727 after take-off from Mexico City, with the resultant fire causing structural failure, in-flight break up and the loss of all on board. The accident investigation revealed that the tire was inflated with air *not* nitrogen as recommended.

Aircraft systems

Modern aircraft are as dependent on the continued operation of their systems as they are on the integrity of the basic structure, for a complete in-flight failure of the electrical or hydraulic systems could prevent them from continuing and completing a safe landing. The vital systems are therefore designed with considerable 'redundancy' so as to be able to continue to operate essential services. They are split into sub-systems to prevent a single failure from having a significant effect on the safety of a flight.

There are four separate hydraulic systems on a B747, each powered by one of the engines, with additional 'redundancy' being arranged by providing an alternative pneumatic power source for each system. With the hydraulic systems each powering separate groups of flight control surfaces, it is considered extremely improbable that the aircraft could ever suffer a complete loss of flight controls. It would require that all four separate hydraulic systems be filled with the wrong type of fluid, or for all the fluid in the four separate systems to be lost, or for some form of explosion or structural failure to take place, for a total failure to occur. Similar care is taken in the design of other systems essential for safe flight.

As the prime sources of power in aircraft are the engines, it is generally true that the level of system redundancy increases with the number of engines, and this is of course a factor in the choice of aircraft types for particular routes. Flights that involve crossing long tracts of ocean, or similarly inhospitable terrain remote from airfields, have traditionally been flown by multi-engine aircraft, usually with four, or perhaps three engines. Operations with two-engined aircraft over these routes normally require substantial modification, including the installation of an auxiliary power unit to provide an acceptable, albeit reduced, level of redundancy.

Minimum equipment lists (MELs)

This term is used to identify specific items of equipment that may be inoperative, or absent, in defined circumstances and yet permit the aircraft to be operated for a flight or series of flights. Aircraft manufacturers produce their own master lists of such items. They cooperate with the authorities and an operator to produce a more restrictive list that is used within that airline by engineers and pilots, in order to determine whether a flight may operate with particular items of equipment inoperative.

The concept is accepted by all affected parties and is described by the FAA in the following terms:

'Experience has proven that the operation of every system or component installed on the aircraft is not necessary, when the remaining operative instruments and equipment provide continued safe operations. Therefore, certain deviations from these requirements are authorised to permit continued or uninterrupted operation of the aircraft.'

Applications of the procedure, however, can cause controversy between engineers and pilots, with the latter group taking the view that the system is abused by some operators to reduce maintenance and operating costs. A traditional philosophy of aircraft engineers is summarised by the statement, 'if it ain't broke, don't fix it', the so-called 'on condition' concept, where items are monitored for performance and do not have a finite life. A response made by pilots to that basic concept, when it results in an accumulation of deferred defects is, 'if it is broke, fix it' and they point out that it is the aircraft commander who is responsible in international law for the safe conduct of a particular flight.

Pilots take the view that a MEL may authorise the operation of a flight with certain equipment inoperative, but the circumstances that apply to that particular flight must be taken into account, e.g. weather, expected operational conditions en-route and at the destination airport, etc. A typical MEL may permit a flight to depart with one fuel contents indicator inoperative, provided that the instruments measuring fuel flow are operative, and that the fuel tank contents are physically measured before departure. A pilot, however, may take the view that if the fuel load for that flight is limited to minimum requirements for some reason, he or she would be justified in refusing to do so. The pilot could have in mind that inoperative fuel contents indicators have been causal factors in accidents.

With airline profit margins under pressure it seems inevitable that managers, engineers and pilots will disagree about MELs and the way they are interpreted, and it is in these circumstances that oversight by the regulatory authority becomes very important.

Extended range twin-engined operations (ETOPS)

The 'rules' that regulated the operation of aircraft remote from suitable alternate aerodromes, such as the Atlantic and Pacific routes, were framed by ICAO many years ago when all types of aircraft suitable for such flights were four-engined. The 'rules' which were not binding on states stated:

'at no point along the intended track the aeroplane is to be more than 90 minutes at normal cruising speed away from an aerodrome ... where it is expected that a safe landing may be made' ...

A further paragraph stated:

'the aeroplane shall be able, in the event of the critical power-unit becoming inoperative at any point along the route or planned diversion therefrom, to continue the flight to an aerodrome ... without flying below the minimum flight altitude at any point'.

This 'rule' was re-examined in the 1980s when large two-engined airplanes capable of flying long distances at lower seat mile costs than three-or four-engined airplanes, became available. Airplane manufacturers and airlines took the view that the old '90 minute rule' should be revoked, but ICAO, the FAA, other regulatory authorities and pilot organisations all believed that the regulation of long-range operations remote from aerodromes by two-engined airplanes was necessary. A long and sometimes bitter series of meetings took place. The position was eventually reached where operations beyond the '90 minute rule' would be permitted after reviews of aircraft designs proved that adequate system redundancy existed, and a verifiable history of aircraft and engine reliability was demonstrated. IFALPA and its member pilot associations continued to express misgivings about the new rules, and the way in which two-engined operations would be monitored by national regulatory authorities. They probably believed that some state administrations would not achieve the high standards required.

Incremental extensions of the 90 minute limitation (60 minutes in the USA) are granted by national authorities to operators of airplane/engine combinations that meet agreed aircraft and system design requirements, have special and restrictive Minimum Equipment Lists (MELs), provide special operational guidance to flight crews and engineers *and* demonstrate required reliability of safety critical systems plus engine shut-down statistics that meet agreed minima until 120 minutes is permitted and then 180 minutes. With 180 minutes, there are very few air routes that cannot be operated, although when some military airfields on the Pacific Islands (Midway, Wake, etc.) are closed as planned, it seems possible that some limitations on route planning will be necessary.

By August 1992, Boeing was able to claim that for the first time more US carrier two-engined airplanes flew the Atlantic than US carrier three- and four-engine airplanes. Also, that in 1993 after 312 000 B767 ETOPS flights, there had been 706 reported events with only 77 events happening during the ETOPS portion of the flights and that of these 77 events only 20 were engine related. (An event is when a pilot must choose whether to divert, turnback or continue.) It seems therefore that ETOPS is a success – perhaps because the initial reluctance of regulatory authorities and IFALPA was overcome by placing tough limitations on these operations – particularly by the increasing redundancy of two-engine airplane systems by in-flight use of auxiliary power units (APUs) etc., and by specifying very low in-flight shut-down rates that must be met by applicant engines.

In the early 1990s new controversy arose over Boeing's application for 'instant' ETOPS for its new airplane, the B777. Boeing claimed that it could demonstrate that the B777 should be authorised for instant ETOPS because of its increased system redundancy design features and improved testing procedures. Over the protests of pilot organisations and some consumer groups, the FAA acceded to these claims. (See Chapter 12 for description of the B777.)

IFALPA complains that there is no central supervision of the collection of in-flight engine shut-down statistics. Also that there are unsubstantiated reports that, on some occasions, engines that should have been shut down were kept operating at minimum power in order to maintain good shut-down statistics. In early 1996, Boeing provided worldwide in-flight engine shut-down data, showing rates on *all* Boeing ETOPS airplane/engine combinations ranging from .002 to .019 shut-downs per thousand engine hours, with .02 as the ETOPS limiting requirement.

At the same time Boeing was suggesting that there is no justification for an airline to have operated an airplane/engine combination for one or two years on non-ETOPS routes before applying to a regulatory authority for ETOPS approval, provided that it has the necessary ETOPS process elements in place. It will seem to some that this suggested further relaxation of requirement will place an added responsibility on state authorities, that some may not be equipped to handle.

An argument pilots use against ETOPS is that Heads of State such as the President of the USA and the Queen of England *never* fly on ETOPS routes airplanes over ETOPS, but this fails to elicit a response!

Continuing airworthiness

The most perfectly designed and manufactured aircraft suffers from stress, corrosion and fatigue from the moment it is pushed out of the factory where it has been built. Throughout its life it is subject to static loads imposed by the loading of fuel, passengers and cargo, with many aircraft suffering more damage from these processes and from collisions with ground vehicles, than they do from all the flights they perform.

When flying, aircraft encounter dynamic loads and stresses imposed by atmospheric turbulence, flight maneuvers and landings, and take-offs. The range of temperatures experienced in flight is as great as 90°C, with the aircraft structure expanding and contracting by several inches while the wing tips may be deflected several feet by atmospheric gusts. Many of the fluids used and some of the cargo carried on aircraft are highly corrosive, and spillages are inevitable, with the alloy skin of an aircraft in the vicinity of the toilets and galleys being particularly vulnerable thus suffering corrosion from leakages of urine and other liquids. A number of aircraft have lost large panels of belly-skin while in flight, and the 'black' museums of airworthiness authorities possess examples of alloy panels so badly corroded that a hole could be made by the pressure exerted by a human finger.

Jet aircraft are pressurised to a differential pressure of eight or nine pounds per square inch, which produces loads of several tons on large doors and windows/exits. It will be seen, therefore, that aircraft constructed to be as light as possible, operating in a hostile environment and subject to stress, vibration, twisting and bending moments while pressurised, carrying dangerous materials, landing at a vertical speed of 700 feet per minute and at a horizontal speed of more than 140 miles per hour, need to have a great deal of expert attention if they are to fly safely for up to fifteen hours a day and 100 000 hours in their lifetime.

Repair, maintenance, overhaul and servicing functions are supervised by airworthiness authorities, performed by airlines or approved maintenance/repair organisations and with advice and spare parts supplied by manufacturers. Airworthiness authorities exercise their supervision by licensing engineers and by the approval of maintenance/repair organisations and of modifications and major repair schemes. Inspection visits are made to ensure that technical standards are at a high level, or alternatively a resident surveyor is assigned to larger organisations.

Engine and aircraft manufacturers keep operators and maintenance organisations informed of technical problems by a system of service bulletins, and occasionally by proposing that specific modifications or repairs be carried out. In the case of larger airlines or organisations, manufacturers have a technical representative in residence. State airworthiness authorities receive all important technical information from the manufacturers, and may decide to make mandatory a repair or modification proposed by them. The standards of quality control achieved by these arrangements are dependent on the performance of the individuals engaged in performing the actual task, or in inspection and approval. It is not surprising, therefore, that human errors causing fatal accidents sometimes occur.

Japan Air Lines B747, 1985

The worst-ever disaster attributable to a failure of engineering occurred on 12 August 1985, when 520 people lost their lives in an accident that occurred in a mountainous region of north of Tokyo when a JAL B747 aircraft was on a flight to Osaka. There were four survivors, one of whom was an off-duty stewardess who provided valuable information to accident investigators and a later inquiry. The crashed aircraft had flown 25 025 hours but because it was used exclusively on short-haul routes it had made 18 830 flights. (The highest-time B747 had flown 62 698 hours and another aircraft of the same type had made 22 970 flights at the time of the accident.)

While flying at an altitude of 24 000 feet en-route to Osaka, the JAL B747 captain contacted Tokyo Control to declare an 'emergency' and request an air traffic clearance to return to Tokyo. Tokyo provided the requested clearance but from the radar plots it was obvious that the aircraft was not flying the course required to return to Tokyo. From those plots, and from information obtained via the aircraft's flight recorders and eyewitness accounts, the following information emerged.

A short period of vibration was followed by an 'event' that caused the nose of the aircraft to dip, followed quickly by the loss of rudder control. Aileron and elevator hydraulic pressure was then lost, and the captain had only differential engine power left to provide a very small measure of control over the inevitably erratic flight path of the aircraft. The aircraft was rolling as much as 40° in either direction, and at the same time was pitching 15° up and 5° down, producing load factors of $0.45\,g$ and $1.85\,g$. While flying over mountainous terrain, the aircraft suddenly turned to the left and plunged into the ground at an angle estimated by an observer to be 45°. The flight time from the 'event' to impact was approximately 44 minutes, and it was a remarkable feat of airmanship for the captain to keep the aircraft flying for that length of time without flight controls. Parts of the aircraft were found 95 miles from the site of the crash and examination of the wreckage revealed that the rear pressure dome ruptured releasing air pressure into the tail of the aircraft. A surge of air pressure literally destroyed the tail control surfaces and caused other damage, resulting in a total loss of hydraulic pressure that in turn caused the loss of all other flight control systems, – statistically a highly improbable event!

Investigation showed that the aircraft had been involved in a landing incident seven years earlier in 1978 that resulted in a 'tail scrape', and that the necessary repairs had been carried out by a team of engineers sent to Japan by the manufacturer. The pressurized bulkhead that subsequently failed causing the disaster, was improperly repaired, weakening the structure with ultimately disastrous consequences.

After investigation of the disaster, attention was given to modifications required to prevent the disastrous consequences that follow the rupture of the rear pressure dome of a B747, and particularly on possible ways to prevent the flow of pressurised air from destroying the tail control surface and causing all hydraulic fluid to be lost from the flight control systems. Boeing decided to offer a modification that fits a sealed metal door over the access hatchway between the fin torsion box and the rear fuselage. Also, to redesign the hydraulic system where it passes through the pressure dome so as to provide more system 'redundancy' and better protection to avoid any further occurrence of the statistically improbable total loss of hydraulic power.

Human factors in engineering/maintenance

Modern civil aircraft are so complex that maintenance work is no longer performed only by licensed aircraft engineers with an inspector responsible for the quality of all work performed. No single person can be competent in all the skills needed to maintain aircraft, engines, systems, and avionics. The work to be done is broken down into specialities, and may take so long to complete that several teams work in shifts for a period of several weeks in order to complete a major overhaul task. This necessarily (but undesirable) complex system of work organisation, sometimes leads to errors that may compromise safety.

Complex aircraft systems may be dismantled by one team, worked on by others, reassembled by another, and be inspected by persons who have not observed the work performed. Parts used have, on occasion been the wrong ones, have been wrongly installed, or have been omitted. Other systems have been adversely affected by work performed, and correct operation of the whole aircraft has not been checked before being returned to service.

Estimates suggest that only 15–20% of workplace errors are caused by the technicians' knowledge, skills, and abilities. The remaining 80–85% are primarily caused by other contributory factors: situation, job and task instructions, task and equipment characteristics, psychological stress and physiological stress. In the USA maintenance organisations are encouraged by the FAA to develop a modified form of 'CRM' for work management.

Engine airworthiness

On 22 August 1985, a twin-jet airliner suffered an engine failure during take-off at Manchester Airport, England. The pilot abandoned the take-off and ordered a passenger evacuation, but a serious fire broke out and 55 lives were lost, mainly by inhalation of toxic gases after the evacuation process had been slowed by difficulty in opening one of the emergency exits, and by the fierceness of the fire along one side of the airplane. Investigation of the accident revealed that a combustion chamber of the left engine had suffered an explosive rupture, uncontained debris penetrated a fuel tank and the extremely large fire that very quickly developed, engulfed the fuselage in only two minutes from the start of the take-off run. The fire spread extremely quickly and passenger evacuation was hindered by flames along one side of the fuselage, due to fuel spillage and the effects of a strong wind. It was found that similar failures of combustion chambers of that engine type had been reported by other airlines, and that the FAA had not ordered mandatory inspection of the subject engine parts. That discovery led to a sharp exchange of views between the NTSB and the FAA on the inspections that should have been carried out in the circumstances.

- In January 1989, a B737 CFM–56 fan failed from fatigue caused by a high-cycle vibration, after a thrust upgrade had been approved without being flight tested and recertificated. Fifty-five lives were lost
- In July 1996, a cracked JT8D–219 fan hub of a Delta MD 88 caused the fan to disintegrate on take-off, penetrating the cabin killing two passengers and causing injury to four

Non-destructive testing (NDT)

In some aircraft accident investigations, serious structural cracks are found that have escaped detection in routine airline maintenance, demonstrating a need for improved inspection methods. As a result, NDT is now of major importance to the airline industry with items such as aircraft wheels being routinely examined at each tire change. In a large airline, as many as 50 wheels a day are examined using NDT methods, with cracks as short as 2.5 mm and areas of corrosion as small as 1.5 mm long and 0.75 mm deep being detected even under layers of paint. Techniques used include: X-ray scanning, infra-red imaging, thermography, dye penetrants, acoustics, and eddy current. In some techniques, the testing processes are visually portrayed on TV screens or stroboscopes, or are digitised and shown on CRTs.

Unfortunately, techniques that deal successfully with crack detection are largely ineffective in detecting corrosion, which needs an NDT technique that can detect material loss. The best one available is radiographic inspection, which remains a difficult technique to apply because of the presence of sealants, paint, adhesives and other protective or water-repellent materials. An increased use of composite structures raises new problems for NDT that have not yet been solved.

A development concurrent with the development of NDT, is the design and manufacture of in-service damage tolerant structures supported by maintenance planning data documents developed for each airplane model. The documents are prepared by the manufacturer in cooperation with the airworthiness authority, and specify maintenance locations, times and intervals. Where possible, NDT techniques are specified to avoid the extremely expensive opening up of structures for inspection, resulting in thousands of man-hours per aircraft inspection.

Responsibility for achieving good results continues to lie with the skill and conscientiousness of operators of NDT equipment, with aircraft fasteners and bolt holes being frequently inspected in situ, and NDT equipment operators being faced with the very difficult problems of access. In some cases, a ground-level scan of a whole airplane by thermal imaging can replace time/consuming X-ray test rigging, but interpretation is difficult in the inspection of complex areas. Thermal imaging techniques have been particularly useful in detecting the ingress of water in honeycomb structures, as differential warming shows several hours after landing.

A major problem is to determine the success rate to be expected of NDT procedures, and the FAA and a UK research center distributed test pieces to both military and civil technicians. In the USA it was found that a 95% success rate of finding cracks could only be achieved if the cracks were 26 mm long and not the hoped-for 8 mm length. UK results were similar.

NDT is essential in the inspection of gas turbine engines, as the materials used are usually working at the limits of their properties with fractured turbine and compressor blades having great potential for causing damage, although they are expected to be contained in cases of failure. Disc failures cannot be contained, and the early detection of cracks in discs is therefore essential if potentially catastrophic accidents are not to occur. Gas turbine engine design makes provision for bore-scope inspection of the engine, and many of these 'hot end' inspections are performed while the engine is on the wing. The equipment used is complex and expensive and demands high standards of skill, but there is little doubt that its use is essential to modern air transport operations in order to provide safety without unduly increasing costs.

The modular design of new engines enables sections of the engine to be changed, and the sections of a particular engine may therefore have experienced greatly different levels of use and be liable to experience different faults in the hours yet to be flown. Only an excellent information system and the use of Supplemental Inspection Documents will permit the required detailed information to be available for the continuous assessment of structural integrity that is required.

Aircraft performance

Regulations that set out specifications for aircraft structures, systems, engines and basic design objectives, have a separate section specifying the required performance and flying qualities of the aircraft. All parts of the regulations have to be considered together, and many variable factors control the performance of an aircraft on a particular occasion.

For the take-off case, variable factors include: the weight of the loaded aircraft, the length and slope of the runway, air temperature, obstructions in the take-off flight path, surface wind, and the possible presence of slush, snow or standing water on the runway. The possibility of engine failure is considered, and flights are arranged so that a pilot is able to abort the take-off and safely come to a stop on the runway, *or* be able to continue the take-off and complete the flight to a safe landing, perhaps at the airport of departure. Should an aircraft suffer engine failure in the en-route phase of a flight, it must be able to over-fly, or fly around, the highest ground en-route and continue to a safe landing, and it must carry the required amounts of fuel and equipment to safely complete the flight in the anticipated conditions. Analysis of rejected take-off accidents by Boeing shows that the decision to reject a take-off is wrong on 66% of occasions, which suggests that information provided to pilots and the stopping performance of aircraft are both in need of improvement.

Pilots know that safety margins available in rejected take-offs are minimal and if the decision to 'reject' is made at $V1$ (decision speed) on a limiting runway, it is almost certain the airplane will overrun the runway. This is because the event may arise for any of several reasons, engine failure, landing gear defects or a system warning for flying controls etc, and the best and quickest decision may take several seconds. During this time two things are happening, the airplane is accelerating and the length of runway remaining is being quickly reduced. Even on a non-limiting runway it is virtually certain the tires will blow and brakes catch fire, so these events are always an incident and sometimes an accident.

For many years USA ALPA made representations to the FAA for increased allowances to be made for these decision times, and although tougher limits are applied to imported airplanes, the 'grandfather rights' syndrome applies to almost all USA built airplanes, and the performance rules current at the time the first airplane of the 'generic family' was certificated apply. These rules assume that the event causing the pilot to reject the take-off and his or her actions to apply all means of obtaining maximum deceleration are simultaneous. Even the techniques used by test pilots to demonstrate rejected take-offs are dubious, as the flight manual states: 1) retard throttles, 2) deploy speed brakes and 3) apply wheel brakes, but test pilots use the technique 1) apply maximum wheel braking, 2) retard throttles and 3) deploy speed brakes. It is interesting that analysis of flight data recorder readouts of rejected take-offs show that the airplane will *always* overshoot the speed at which the pilot applied wheel brakes.

The reason for continuing disagreement about rejected take-off rules is that the 'old' rules allow more payload to be carried from limiting runways, and manufacturers and airlines resist changes, although the FAA has issued proposals for rule changes.

Flying qualities

The subject of flying qualities has become more complicated by a greater use of automatic flight control systems. Where it was once sufficient to specify the maximum forces required to operate the flight controls and to add a general requirement that the aircraft should be capable of being flown by pilots of 'average' ability, it is now necessary to take account of the effects of malfunctioning autopilots and powered control and trimming systems.

This is achieved by designing redundancy into the systems and arranging automatic and 'graceful' degradation through several stages until a minimum 'adequate to get you home stage' is reached. A major advantage of automatic, fly by wire, fully-powered flight control systems is that software changes can produce almost any desired flying qualities so that 'feel', harmonisation and control forces can be optimum. Where a family of airplanes is built, B757/767 and Airbus A330/340, each type in a family can be provided with near-identical flying qualities so easing pilot qualification for customer airlines.

During B707-300 test flying for the purpose of obtaining British certification, it was required to demonstrate 'minimum unstick speed' (V_{mu}). This was because of earlier accidents to two Comet jetliners caused by pilot attempts to take-off at too low a speed with the nose of the aircraft raised so high that the tail was touching, or almost touching, the runway, causing the resultant very high 'drag' to exceed the thrust available. Both Comet aircraft failed to become airborne with fatal consequences, and special testing of the B707 was required to ensure that it did not suffer from the same deficiencies of flying qualities and performance during take-off as had the Comet.

For the purpose of safely performing this test demonstration, the longest runway available over the dry lake bed at the Edwards Air Force Base in California was used. Anticipating that the test might produce unusual results, a large baulk of timber was fixed under the fuselage near the tail to protect the aircaft structure, and this proved to be a very wise precaution. The test pilots applied up-elevator from an early stage of the take-off run and the timber was soon in contact with the surface of the runway. Photographs of the dramatic and unsuccessful attempt to take-off (see Figure 3.5) show the aircraft using miles of runway with the timber in contact with the runway surface. Heat produced by friction caused the timber to smolder and it was clear that the aircraft could not take off. As a direct result of the tests, that model of the B707 acquired a ventral fin below the fuselage and near the tail, changing the ground geometry, making it impossible to over-rotate the aircraft during take-

Figure 3.5 Certification test flight – BOAC B707.

off, and so providing acceptable performance and flying qualities.

There have been many occasions when for a variety of reasons, aircraft have been flown into a situation where good flying qualities were needed to prevent an incident from becoming an accident. Pilots have inadvertently stalled large passenger aircraft for a variety of reasons, such as flying at too great an altitude for the operating weight, or levelling off after a descent performed by the autopilot and then failing to restore engine power.

In 1985 a Boeing 747, flying at 41 000 feet, 300 miles northwest of San Francisco, lost thrust from number four engine. The crew became pre-occupied with restoring power to that engine, the pilot failed to monitor his flight instruments and left the autopilot engaged. When he disengaged the autopilot he did not re-trim the aircraft for an asymmetric power condition. The aircraft rolled right, nosed over and entered an uncontrolled descent through cloud during which the crew became disorientated. Control of the aircraft was regained after descending to 9500 feet when visual flight conditions were experienced.

During the period that the aircraft was out of control, it dived at an angle of 67° and rolled through 176° to an inverted attitude. The aircraft exceeded all airspeed limitations and incurred 4.5 g during the recovery maneuvers, causing the landing gear up-locks to be dislodged and the gear to fall through the landing gear doors. The left outboard elevators and three-quarters of the right outboard elevator separated from the aircraft, the horizontal stabilisers were severely damaged and all fluid was lost from number one hydraulic system. It is a great tribute to the strength of the aircraft that it did not break up during the dive and to its flying qualities that the pilot was able to recover from the 'upset' and fly the aircraft to a safe landing at San Francisco (see Figure 3.6).

A particular problem caused by the need to operate jet-powered aircraft at high altitudes in order to reduce fuel consumption, is the so-called 'coffin corner' that is encountered if a pilot should attempt to fly the airplane at too high an altitude for the existing operating weight. For any airplane a speed envelope can be drawn. The left side is V_s, the 1 g stall speed, which bends to the right with increasing altitude as a result of Mach effect. To the right side of the envelope is V_{mo}/M_{mo}, the maximum operating speed boundary determined by the structural strength requirements to limit operating speeds so that the airplane may safely encounter gusts (and by test flights that examine flying qualities), which bends to the left to meet the 1 g stall speed boundary at 'coffin corner'. An attempt to operate an airplane at the altitude and speed designated by the intersection of these boundaries can be likened to attempting to balance on a pin head. Flying one knot slower will stall the airplane, and

Figure 3.6 Boeing B747 at San Francisco (*Courtesy of NTSB*).

flying one knot faster will put the airplane into high speed buffet. The airplane has no reserve of flying qualities or performance and an encounter with a gust would result in serious handling problems, perhaps leading to loss of control (see Figure 3.1).

Future problems – flying qualities

Each new aircraft type entering service is required to be more economic in operation than types replaced. A design technique used to achieve this purpose is to reduce the basic stability of the aircraft so reducing the area of tail surface required, the weight and drag of the aircraft, and power and fuel used. Natural stability may eventually be so reduced that the human pilot will not be able to safely control the airplane by direct use of primary flight controls, and pilot control inputs will be accepted and interpreted by digital computers in the automatic flight control system. This will be the first time in civil aviation that complete authority has been given to such systems.

Total in-flight failure of the systems would have serious consequences for flight safety, and airworthiness authorities will demand that the systems have a higher degree of proven reliability than any other aircraft system. The use of digital computers and other basic design considerations will make it essential to establish the initial and continuing airworthiness of computer 'software' and it is not yet clear how this can be assured. Experience gained in military aircraft will be valuable in designing and proving the systems, but acceptability of the concept for military missions cannot be automatically transferred to civil aviation.

Flight control computers and electrical power supplies must be provided with complete protection from electro-magnetic interference and power source interruption, for it would not be acceptable for a lightning strike to threaten performance of the flight control system. An increased use of new composite and non-metallic materials in aircraft structures adds a further complicating factor to ensure that lightning and electromagnetic interference cannot adversely affect the flying qualities of these new types of aircraft. At least one military airplane has been destroyed because it flew within reception range of radio transmissions that disabled its 'fly by wire' control systems, and a very careful evaluation of the environment in which these types of aircraft are operated is essential.

Helicopter airworthiness

The generic term 'aircraft' includes airplanes, gliders, rotary wing aircraft embracing autogyros and helicopters, and lighter-than-air machines such as airships and balloons. However, only airplanes and helicopters are of commercial significance, and receive serious attention from the International Civil Aviation Organisation. There are many more fixed-wing airplanes in commercial service than helicopters, but the unique abilities of helicopters command attention as they are in wide use for the performance of special tasks beyond the competence of airplanes. These tasks include the carriage of passengers and high value cargo to and from difficult and remote sites such as offshore oil rigs, mountain tops, city centers, ships in distress, and less frequently to ships at sea. They are also used to provide links between airports, and between airports and city centers, and for power line inspection, police work and crop dusting. All serious studies of air transport include the helicopter, and its special qualities have caused almost all development to be performed by the military, although the general public might see the case for further development being its worldwide use for the humanitarian task of search and rescue.

The ability to perform vertical flight and to hover, requires the helicopter to obtain almost all of its lift from rotating wings, and this fundamental difference between the airplane and the helicopter gives rise to the helicopter's advantages and disadvantages compared with the airplane. Large helicopters use much of the technology of airplanes, with gas turbine engines, structure, systems and instruments sharing many features, but the need to rotate the wings adds a difficult problem to the task of the helicopter designer. The mechanical system used to transmit power from the engine(s) to the rotor(s) is extremely complex, and is a prime cause of high levels of stress, vibration, fatigue and costs, with lower levels of reliability and safety. In the case of the largest civil helicopters, two large rotors are driven from a common shaft that is itself powered by geared drives from two large gas turbine engines. The mechanical gear train has to be capable of powering both rotors from one engine in the event of engine failure, and to ensure that the rotors remain in synchronisation because their paths overlap. This complexity poses a severe design problem and rotors and gear trains create much work for maintenance/engineering organisations.

When a hostile environment is added, such as low-level operations over the sea to and from offshore oil fields, corrosion, stress and vibration combine together to be the causal factor(s) of many helicopter incidents and accidents. Because of the extremely strong political and economic case for the use of helicopters in offshore work, there has been a tendency to write airworthiness and operating rules to match existing capabilities of the vehicle rather than (as in the case with fixed wing aircraft) to write rigid requirements and demand compliance. A combination of relaxed airworthiness and operating rules has been a major cause of the accident rate of helicopters being about five times worse than fixed-wing aircraft. However, comparisons are extremely difficult to make because of the dearth of compatible statistics, perhaps because few helicopters are operated in a public transport role.

One area where the helicopter is markedly inferior to fixed-wing aircraft, is its ability to operate in icing conditions. This deficiency poses both economic and safety problems for helicopter operators, with the main threat to helicopter safety coming from the deterioration of the main rotor's aerodynamic characteristics. In severe icing conditions rotor imbalance is observed, for the rate at which ice is formed on rotors varies with local airspeed, increasing with rotor rpm and radius from the hub. Kinetic heating also has a marked effect, and the amount of ice varies considerably with rotor and aircraft speed. Severe imbalance can occur and it is not unusual for helicopter operations to be seriously curtailed in weather that is a mere inconvenience for fixed-wing aircraft.

Helicopters are inherently unstable, and this characteristic gives rise to many problems in the fields of control and human factors. Piloting helicopters is a very demanding task, and fatigue arises because of that factor and because of the very large number of flight sectors flown in a given number of hours, many of them at low level and in poor weather. Noise levels and vibration are all much higher than in fixed-wing operations and justify much more severe and protective regulations being enforced than has yet been the case.

Helicopter safety

In 1990, ICAO sought to restrict operations by single-engined helicopters by banning them from use in instrument conditions at night, out of sight of land, with cloud ceilings below 600 feet in visibility of less than 1500 metres (4900 feet) and from elevated heliports. This was strongly resisted by manufacturers and operators who pointed out that these helicopters are 75% of the operating fleet. As member states may choose *not* to apply ICAO rules, it is possible that it will be some years before these restrictions apply on a worldwide basis. (The USA filed 54 'differences' from the new ICAO rules).

Safety statistics for helicopters are difficult to obtain in a form that makes safety assessments and comparisons possible. In 1993, the UK Civil Aviation Authority dubbed the UK *fatal* accident rate as 'very worrying' because the three year moving average increased from less than 1:100 000 hours in 1987, to more than 3:100 000 hours in 1991. A three year moving average is used because the small numbers of helicopters in use in the UK mean that annual figures greatly distort the picture. At the same time the Helicopter Association International reported that civil helicopters in the USA had improved their safety record from 30 accidents per 100 000 flying hours in 1970, to about 12 per 100 000 flying hours in 1990, and 10 in 1994. USA figures are derived from a much larger data base than the UK, as in 1994 the USA had 4752 helicopters and the UK only 365.

A great deal of progress has to be made before helicopters improve their worldwide fatal accident rate from about 30 per 10 million flights to the one per 10 million flights of jet airliners! A notable exception to the generally poor safety records of helicopters is that of a US operator, Petroleum Helicopters Inc., that achieved a three year accident-free record of 500 000 flight hours, and more than two million take-offs and landings after launching a safety incentive program for employees.

A major factor in reducing helicopter accident rates is the widespread adoption of Health and Usage Monitoring Systems (HUMS). Here, flight data, engine and gearbox parameters from existing sensors and vibration data from specially installed accelerometers are fed into an on-board computer. This analyses, filters and reduces data to manageable quantities that can be stored in the memory and downloaded on landing to a ground-based computer. Together with analyses of gearbox oil and debris, and the collection of data on rotor track balance, these techniques provide an effective means of monitoring helicopter condition and predicting failures.

Aircraft cabins – crashworthiness

Experience dictates that consideration be given during the design stage of an aircraft project, to mitigate the worst consequences to passengers and crew should a crash occur during take-off or landing. All items contained in the cabin of airline aircraft are studied to ensure that every possible precaution has been taken to minimise risk arising from incidents and accidents.

Passenger seats are designed and stressed to stay in place except during the most severe accidents, so providing the best chance of survival for an occupant restrained by a seat belt. A long-running safety debate continues on whether passenger seats should face the rear of the aircraft to improve chances of survival in cases where high deceleration forces are experienced Expert opinion is almost unanimous that the chances of surviving an accident would be increased should rearward facing seats be fitted. However, airlines resist change for reasons of increased costs and alleged passenger preference, although the USA Air Force requires that all seats fitted to its military transport aircraft be rearward facing and stressed to 16 *g*.

Pilots are required to wear full safety harness during take-off and landing, for it is considered that they must at all times be able to manipulate controls and manage the aircraft systems. For example, the most important emergency actions pilots perform in an emergency landing are to close down the engines and shut off the fuel so reducing risk of fire. The controls for these systems must remain within their reach even after severe decelerations. Aircraft equipment is required to be restrained, and flight attendants are responsible for safely stowing all loose equipment before take-off and landing. The amount of 'duty-free' goods carried on aircraft has caused problems during emergency evacuations. Flight and cabin crews are united in a belief that it is ridiculous for passengers to make purchases in the duty-free shop and then carry these inflammable items by air to their point of destination, when it could be arranged for them to buy the same product at the airport of arrival.

Some airlines fail to exercise sufficient control over the amount of cabin baggage passengers are allowed to carry aboard, and flight attendants are placed in a difficult position by their employers. They must then appeal to the authorities to impose tougher requirements on operators, perhaps by having Flight Operations Inspectors make unannounced inspections of aircraft cabins or observe the check-in procedures used. Aircraft designers have provided some relief for hard pressed flight attendants by installing larger overhead storage bins for passenger use, but doubts are expressed about the ability of the bins to remain secure in emergency conditions. One passenger won a $1 million lawsuit against an airline when a bag fell on his head from an overhead locker.

It is essential that emergency exits and aisles of aircraft are sufficiently wide for their purpose and are free of obstructions if an emergency evacuation is to succeed in the worst case. Ease of access is critical if the target of 90 seconds for a complete evacuation of an aircraft is to be achieved. Lessons that were learned from fatal accidents caused improved access to emergency exits to be provided, and for floor-level lighting to be installed so as to improve passenger evacuation rates in dark or smoke filled cabins.

4: The Natural Environment

In the earliest days of aviation it was not possible to fly safely except when the weather provided good visibility and light winds. However, military necessity and commercial pressures combined to provide incentives to operate in 'all weather' conditions. Vast sums of money have been expended in research programs and in learning to fly safely in almost any weather, and modern airliners operate when the cloud base is zero, visibility is 50 meters and the runway is wet from various forms of precipitation. Cross winds of 30 knots or more are acceptable for landing and the rare occasions when airports are closed are usually due to near-zero visibility, hurricane force winds, or the runway surface being unacceptable for reasons of great depths of standing water, snow or ice. Although progress made in providing regularity to airline operations is impressive, natural phenomena exist that persuade experienced pilots to delay flights, and these phenomena cause accidents to aircraft piloted by less experienced or less cautious pilots.

Winds and turbulence

Winds of great strength such as those found in hurricanes/typhoons, and tornados/cyclones may create turbulence sufficient to threaten the structural integrity of transport aircraft. They are usually forecast and plotted by satellite or weather reconnaissance aircraft, or can be detected by weather radar equipment carried on airline aircraft. It is therefore unusual for airliners to experience close encounters with these phenomena, as pilots make aerial detours of many miles, or delay the flight, to avoid the hazard. Clear air turbulence (CAT), associated with strong upper-level winds known as jet streams, can cause turbulence sufficient to create passenger discomfort and fear, but has not, so far as is known, caused the loss of any aircraft although it undoubtedly uses up aircraft fatigue life. Research is being conducted to learn more about

CAT, with the hope of developing equipment that will provide detection in much the same way as weather radar enables pilots to detect the heavy precipitation usually associated with the weather systems that cause normal air turbulence.

Until such systems are available, meteorologists provide reasonably accurate forecasts of the location of CAT and pilots plan routes accordingly, but surprises cannot be ruled out. In 1995, 26 passengers and crew were injured on an American Airlines A300 airplane near San Juan, Puerto Rico, and in the same area on a different date 20 passengers and crew were injured aboard a Continental Airlines airplane. In similar incidents in 1990, very strong winds over the mountains of Iran caused two airplanes to report encounters with standing waves, with a B747 losing 4500 feet in severe turbulence although full engine power was being used, and a following airplane forced into 2000 feet per minute climbs and descents. The wind speed was reported to be between 140 and 180 knots.

A more dangerously violent form of turbulence is found in the rotor-like flow in the immediate lee of mountain peaks, when very strong winds flow over mountain ranges. This rotor turbulence can be severe to a degree sufficient to destroy any transport aircraft. In the 1960s a B707 departing from Tokyo Airport on a fine clear day diverted from the direct track to Hong Kong, with the approval of air traffic control, for the purpose of showing the beautiful Mount Fuji to the passengers. The aircraft flew into a rotor flow in the lee of Mount Fuji and broke-up in extremely severe turbulence, causing the death of all on board. The lessons learned from this accident were widely publicised and a similar accident should not occur again.

Heavy precipitation

Thunderstorms and well developed cumulus clouds produce heavy precipitation at all levels, in the form of hail or very heavy rain. In tropical latitudes hailstones may reach the size of golf balls, and aircraft that encounter such hail are severely damaged, particularly on the nose, leading edges of wings and tail surfaces and engine cowlings. The ingestion of large hailstones can cause severe damage to fan blades of engines, and pilots learn to respect such phenomena and to use weather radar to avoid encounters whenever possible. When an encounter is made, damage may run to millions of dollars and lead to aircraft being grounded for very expensive repairs that can include the need to re-skin part of the wing, fuselage or tail. In some parts of the world, thunderstorms are so frequent and widespread that penetration by airliners is inevitable, but in these cases a prudent pilot will slow the aircraft to a safe speed so as to minimise damage and will ensure that occupants use seat belts and loose articles are stowed.

Very heavy rain is less dangerous to flight although there is evidence to suggest that the heaviest rainfall adversely affects aircraft performance, which has consequences for safety during take-off and landing but not normally during cruising flight. The most serious consequences of heavy rainfall are experienced during landing and take-off, when flooded runways cause damage to wing flaps and engines (the latter through ingestion of water), and even more seriously adversely affect aircraft performance by extending the take-off run. In the landing case, pools of water on the runway may cause the aircraft to aquaplane and over-run the paved surface because of an absence of effective wheel braking in these conditions, or a loss of directional control can cause the aircraft to exit the side of the runway. These risks are minimised by good design of runways and the use of porous top surfaces, but in some cases a pilot is well advised to delay take-off or landing until excess water has drained from the runway surface. When flooding does occur on runways, tires can be damaged by water trapped between the tire and the runway turning to super-heated steam.

In 1992, Boeing warned all operators of CFM 56-3 powered B737s that there had been three dual engine flame-outs and one single engine flame-out on that airplane type over a two year period, and that modifications and changes to operating procedures were necessary. In moderate or severe precipitation, continuous ignition (both igniters) and a minimum engine speed of 45% are required plus, in severe turbulence, switching off of the auto-throttle systems. A system to warn of low engine speed was installed.

Heavy rain obscures pilot vision through the windshields, and in a worst case can cause optical illusions from refraction and reflections that are particularly troublesome at night. To counter these effects, specially formulated rain repellent liquids are sprayed onto windshields in the worst conditions of very heavy rain. The liquid used is difficult to handle and is not environmentally friendly, and a welcome later development is the use of hydrophobic coatings on windshields that are equally effective. These coated windshields are standard on all new-build Boeing airplanes.

Aircraft icing

Engine and airplane icing have caused accidents by adding weight, and/ or reducing lift or engine power since the start of aviation, and air and ground systems are used to minimise risk. On the ground, airplanes are de-iced by fluid spray systems which can be effective for as long as 30 minutes, depending on the fluids used and the meteorological conditions. In flight, wings and tailplanes are de-iced by heat or inflatable boots and engine systems are de-iced and/or anti-iced by hot air as required. Some airplane types are more vulnerable to icing risks than others due to design features or operating conditions. Some models of DC9s and MD80s were prone to accumulate wing ice due to cold-soaked fuel reducing the temperature of wing surfaces so that ice formed while the airplane was on the ground. Some turbo-prop airplanes were found to be susceptible to wing ice formation in flight that caused aerodynamic aileron-lock leading to a loss of control. Other airplane types have suffered ice-induced tailplane stalls leading to loss of control. In some of these cases, use of wing flaps exacerbated control difficulties.

In all these cases, special modification action was taken or new rules governing flight in icing conditions were applied. These newly severe FAA rules prohibiting take-off with ice contaminated aerodynamic surfaces is leading to the design of ice detection systems that will remove guesswork from pre-flight inspections of airplanes on dark nights. There have been calls for mandatory inclusion of improved flight test examination of handling qualities in severe icing conditions. Recent research has shown that flight in freezing drizzle when small water droplets coalesce may produce severe icing conditions similar to those found in freezing rain, and so it is likely that icing will be an unending threat to air safety.

Accidents due to aircraft icing continue to occur. One of the worst in recent years happened to a Boeing 737 taking off at Washington Airport in January 1982, when it crashed into the River Potomac leaving only five survivors from the 79 persons on board. The accident inquiry determined that the cause of this accident was airframe and engine icing that impaired the performance of the aircraft. This caused false indications of engine power to a degree that misled the pilot into believing that the take-off was normal, until too late to become safely airborne or to discontinue the take-off. Other fatal accidents attributed to icing were to a DC9 at Cleveland in 1991, and an ATR 72 turbo-prop at Chicago in 1994.

Slippery runways

The annual onset of wintry conditions of sleet, snow and icy runways is the reason for precautions taken by experienced airlines. They arrange to have their flight and ground crews review, prior to the start of the winter season, problems associated with operations conducted from runways affected with snow, ice or slush. Many serious accidents have been caused by aircraft landing on slippery runways and leaving the paved surface, either at the end or sides, while still travelling at speed. In March 1976, a Japan Air Lines Boeing 747 at Anchorage Airport, Alaska was blown *backwards* off an icy taxiway into a ditch, causing severe damage to the aircraft but no serious injuries. Other accidents have been caused by aircraft failing to gain flying speed when taking-off from slush-covered runways.

In recent years there has been an increased emphasis on the early removal of snow, ice, slush or standing water from runway surfaces. The design of runways includes the provision of surfaces that are capable of providing good coefficients of friction over a long life, together with good water drainage characteristics. The removal of rubber deposited on runways by aircraft tires to restore good friction characteristics is an annual task for maintenance crews at better maintained airports. These measures, when adopted, help maintain airport safety. Each winter there are accidents on slippery runways, athough efforts are made to provide pilots with information on runway conditions by use of special friction measuring machines that are towed along the runway.

Windshear

A dangerous form of severe weather is a near invisible hazard: the microburst (see Figure 4.1). This causes windshear that may be so severe that it will force to the ground any aircraft that experiences an encounter in the most adverse conditions (see Figure 4.2). Meteorologists describe microburst windshear as a downburst, or strong downdraft, approaching the ground and inducing an outflow of potentially dangerous winds and downflow shear at low altitude. If the area of the downburst is confined to no more than two nautical miles in area, with peak gusts lasting from one to five minutes, it is classified as a microburst. A larger downburst with damaging winds lasting from five to twenty minutes is classified as a macroburst.

A natural phenomena, windshear has always been a hazard to safe flight. Its nature and severity only became recognised and explained during the late 1970s and early 1980s after accident investigators searched for causal factors for accidents that clashed with conventional theory about aircraft performance in the take-off and landing config-uration. Digital flight recorders provided evidence to corroborate the

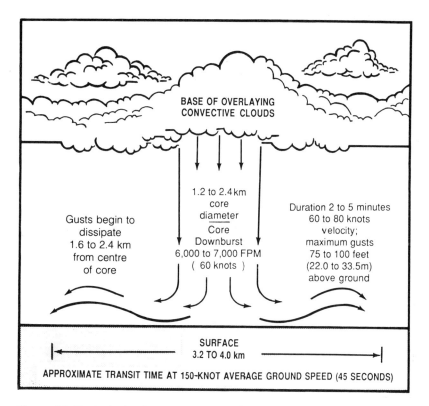

Figure 4.1 Typical microburst cross-section (*Courtesy of ICAO Bulletin*).

testimony of pilots who had survived what were regarded as inexplicable accidents. It is accepted by some safety analysts that the quest of airplane designers for shorter take-off and landing distances has reduced aircraft performance in the take-off and landing phases of flight to a degree that makes modern jet-powered airplanes more susceptible to windshear than earlier types.

Windshear often occurs in visible and violent weather conditions such as thunderstorms. It can also occur in hot dry climates and in other circumstances where its existence may be more difficult to detect. The effects of severe windshear on aircraft performance are abrupt changes in airspeed and groundspeed, with consequential and significant effects on airplane performance. Should the airplane be flying at low speed, and in a configuration appropriate for landing or take-off, airspeed and height may be lost to a degree that causes premature contact with the ground, notwithstanding the best efforts of the pilot

In the USA, there were 14 air carrier windshear-related accidents from 1975–1985 causing more than 400 deaths, and there have been others

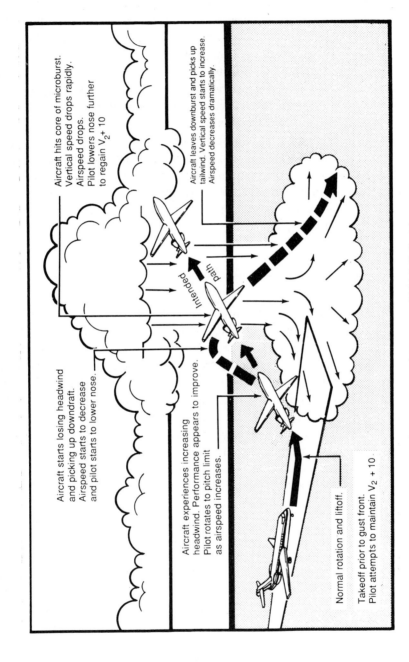

Aircraft hits core of microburst.
Vertical speed drops rapidly.
Airspeed drops.
Pilot lowers nose further
to regain $V_2 + 10$

Aircraft leaves downburst and picks up
tailwind. Vertical speed starts to increase.
Airspeed decreases dramatically.

Aircraft starts losing headwind
and picking up downdraft.
Airspeed starts to decrease
and pilot starts to lower nose.

Aircraft experiences increasing
headwind. Performance appears to improve.
Pilot rotates to pitch limit
as airspeed increases.

Intended path

Normal rotation and liftoff.

Takeoff prior to gust front.
Pilot attempts to maintain $V_2 + 10$.

Figure 4.2 Take-off profile in a microburst situation (*Courtesy of ICAO Bulletin*).

since then. The FAA, Boeing, FSF and ICAO worked together to produce a windshear training aid that is in wide use, and many airlines program windshear events into flight simulators. In the worst windshear case recorded in July 1989, a Boeing 737 experienced a 95 knot headwind loss at Denver's Stapleton Airport, but the pilots were able to avoid an accident because of their training.

Efforts have been made to design predictive-type windshear detection equipment using doppler radar, infra red and lidar (light detection and ranging) systems. Manufacturers are now producing and selling operational systems (mainly based on doppler radar) that can provide up to 90 seconds warning of windshear events ahead of the airplane. By early 1996, three forward-looking windshear radars had been certificated in the USA. Together with installations of ground-based low level windshear alert systems that detect significant windshear from an array of anemometers installed in the airport area, good progress is being made in this safety program.

Lightning strikes

A lightning strike on an aircraft in flight is unpleasant for pilots and causes alarm among passengers. A number of fatal accidents have been caused by strikes, but these are now less frequent since aircraft designers learned to provide good electrical bonding of the aircraft and to arrange special protection for fuel tank vents. In the 1970s, a Boeing 747 of the Iranian Air Force was destroyed by a lightning strike while in flight over Spain. A civil Boeing 707 was similarly destroyed over the USA a few years earlier. There have been other cases of severe damage to aircraft struck by lightning, although the metal skin and structure of conventional aircraft form a Faraday Cage which normally protects the crew, passengers and equipment. However, new problems are arising as composite and non-metallic aircraft parts come into use.

The huge number of lightning strikes that occur on a daily basis around the world, bring the relatively few instances of destruction and damage to aircraft into perspective. There are approximately 1800 storms in progress at any given moment, producing 44 000 thunderstorms, which generate nine million lightning flashes each day. When weather satellites and lightning flash counters came into use, the magnitude of the earth's electrical activity became better known. It is now appreciated that sand and dust storms, snow storms, volcanic eruptions and man-made atomic explosions can also generate lightning.

A major safeguard against lightning strikes, is the use of weather radar in airline aircraft. Lightning is frequently associated with thunderstorms, and these can be detected and avoided owing to the precipitation detection capabilities of the radars. It is possible to detect lightning discharges where no precipitation is present, by use of a passive electrical signal reception device giving a display on a cathode ray tube (CRT), or on a liquid crystal display (LCD). Ranges of 200, 100, 50 or 25 nautical miles can be selected by the pilot, but few of these devices are installed in airline aircraft, as of 1996.

Volcanic eruptions

There are a large number of volcanic eruptions each year, but they received little attention from civil aviation until the 1980s, when a number of airlines reported that they were having to replace cockpit windscreens and passenger cabin windows on a scale previously unknown. It required a procedure akin to detective work to discover that the worst cases had a common factor. The affected aircraft were all assigned to air routes in the most northerly latitudes of the northern hemisphere. The damage was being caused by flight through dust and ash thrown into the earth's atmosphere by an eruption of the volcano El Chichon in Mexico, in the months of March/April 1982. The millions of tons of ash and sulphur dioxide thrown into the upper levels of the earth's atmosphere by the eruption, were distributed by upper-level winds (the jet stream) throughout the northern hemisphere, with the greatest concentration in northerly latitudes. There were other major eruptions over a relatively short period of time, including Mount St Helens in May 1980, one in the Marianas in March 1981 and another in the Kurile Islands in April 1981, all of which added to the effects of El Chichon.

The aircraft windows and windscreens most affected were acrylic-covered, for this material is a poor electrical conductor, becoming charged with static electricity from contact with the ash and sulphur. The static electricity generated, attracted even more particles onto the windows causing a scouring effect that badly affected their visual properties. A further adverse effect arose with atmospheric water vapour added to the particles of sulphur, and the 'mix' becoming a corrosive sulphuric acid. Palliatives were adopted by aircraft and window manufacturers and airlines, and the matter was seen as being of an economic nature and not a safety related problem.

That perception would not hold for the results of a volcanic eruption of Mount Gallunggung, Indonesia, that occurred on 24 June 1982, when

ingestion of volcanic dust caused all four engines of a British Airways
B747 to fail.

British Airways Flight 009

Captain Moody was in command of a B747 aircraft when it departed
Kuala Lumpur for Perth, Western Australia with 247 passengers and 16
crew members on board for what was expected to be a routine flight.
When established in cruise at 37 000 feet about 130 nautical miles south
east of Jakarta, Indonesia, Captain Moody visited passengers in the first-
class cabin but was called back to the flight deck by the copilot. On
climbing the stairs to the upper deck, he noticed what appeared to be
puffs of smoke from the floor-level air conditioning vents and an acrid
smell. Windshields were aglow with what appeared to be a brilliant
display of St Elmo's fire, and the copilot and flight engineer had already
switched on passenger seat-belt signs and engine igniter systems. On
checking the weather radar, Captain Moody noted that nothing of
significance was displayed on the screen. The copilot then pointed out
that the engine intakes were glowing as if illuminated from within, and
that an unusual stroboscopic effect was making the fans of the engines
appear to be slowly turning backwards. The display of St Elmo's Fire on
the windscreen turned into a display similar to tracer bullets, and at
about that time the flight engineer called out: 'Engine failure number
four!'

The crew carried out the engine fire drill, and then all three remaining
engines ran down, leaving Captain Moody with the statistically
improbable occurrence of a modern jet aircraft with four failed engines.
An emergency distress call was made and emergency drills for four failed
engines were carried out. Unsuccessful attempts were made to re-start
the engines as the aircraft descended, and at 26 000 feet the cabin
pressure warning horn sounded because the engines were not delivering
an air supply and the cabin pressurisation was being lost. The flight crew
put on oxygen masks as they had done many times before in flight
simulator training sessions, but on this occasion the copilot's mask came
to pieces in his hand, causing yet one more element of emergency; at that
altitude the copilot would quickly lose consciousness without oxygen.

Faced with a difficult choice of conserving altitude in order to prolong
the flight of his giant-sized glider, or lose the sorely needed services of
one of his crew, Captain Moody elected to increase the rate of descent of
the aircraft, turning towards the sea and away from high ground. By the
time the aircraft reached 20 000 feet, the copilot had reassembled his
oxygen mask, and a reduced rate of descent was again possible. The

inertial navigation systems were displaying 'garbage', and the airspeed indications shown to the pilots disagreed by 50 knots. At about the same time, the cabin altitude reached 14 000 feet, passenger oxygen masks were automatically released for use and Captain Moody decided that it was time to provide the passengers with information about their plight.

> 'Good evening ladies and gentlemen, this is your captain speaking. We have a small problem. All four engines have stopped. We are doing our damnedest to get them going again. I trust you are not in too much distress.'

The crew succeeded in restarting number four engine which was the first to run down, and 90 seconds later the other three engines were re-started. Captain Moody was able to initiate the climb of the aircraft from the dangerously low altitude of 12 000 feet that had been reached, and to head towards Jakarta. When the aircraft reached an altitude of 15 000 feet in the climb, St Elmo's Fire reappeared and number two engine began to surge, causing Captain Moody to conclude that all four engines might have been severely damaged. On reaching Jakarta, Captain Moody had great difficulty in seeing the runway lights and realised that the windows and windscreens were opaque. Although he could see very little, he made a safe landing to loud cheers from his passengers, most of whom had not expected to survive this amazing incident.

On inspecting the aircraft, it was seen that the leading edges of the wing and tail surfaces, engine nacelles and nose cone were stripped of paint as if the aircraft had been sand blasted. The engine turbine blades were badly damaged and partly worn away, and it was confirmed after two days of intensive investigation that the damage and engine failures had been caused by flight through a thick cloud of volcanic dust thrown up into the earth's atmosphere by the eruption of Mount Gallunggung.

Since that date, there have been other incidents arising from airliner encounters with volcanic eruptions, including a KLM B747 – 400 having all four engines flame-out over Alaska in 1989, and a system of observing and reporting these natural phenomena was introduced. The first efforts were made in Australasia, but the system is now worldwide in application, with eruptions observed and debris tracked over thousands of miles by land-based observers or by satellite systems. Pilots are warned at pre-flight briefings or, if necessary, by air traffic services while in-flight.

Visibility

Sandstorms, dust, snow, heavy rain and fog, restrict surface visibility at airports and are factors adverse to safe and regular air transport. It took many years to develop instrument flying techniques and equipment for the purposes of freeing airline operations from the worst effects of poor visibility. Accidents were caused by pilots having unexpected encounters with fog at airports, and forecasting the problem was extremely difficult. Small changes of wind speed or temperature are sufficient to make fog form or clear, if there is a general disposition to that condition. The worst forms of fog from a pilot's point of view are those that do not have a consistent density but have dense patches within the less dense general fog. It is unusual for a year to pass without at least one accident with fog as a causal factor.

Many accidents resulted from attempted landings in reduced visibility, and some were caused by the pilot suddenly losing all visual references on entering a shallow layer of fog just before touchdown. There have been many occasions when airline operations have come to a standstill because of widespread fog at airports. A method of landing and taking-off in reduced visibility was early recognised as a high priority for the purpose of maintaining safety in these conditions, and in reducing costs incurred by the disruption of airline schedules.

Attempts have been made to restore good visual references by dispersing fog. During World War II, many military airplanes were landed safely at fog-bound airfields, because heat produced by large gasoline burners set alongside the runway dispersed the fog by raising local temperatures above the temperature (dew point) needed to sustain it. However, costs incurred by this method were too great to be accepted by civil aviation. Other methods were therefore devised to enable flight operations to continue in low visibility (see Chapter 12).

5: The Operational Environment

The international nature of the civil air transport industry requires the cooperation of many states to provide an acceptable operational environment. However, because the levels of technical expertise and financial resources available to states vary greatly, standards of: air traffic control systems, airports, communications, security, emergency procedures and equipment, aerodrome ground aids, navigational facilities and weather forecasting, also vary. The United Nations Development and ICAO Technical Cooperation Programs have funded significant improvements in recent years. These initiatives are aimed at removing the worst deficiencies in the operational environment, particularly in 'developing world' countries. At the same time, a number of more affluent countries made additional contributions of funds, equipment and advice to needy countries on a bilateral basis. Some of these national programs are flawed by being tied to the technological products of the donor country, and occasionally result in the recipient country receiving equipment not matched to its needs or beyond its ability to use and maintain.

Air traffic services

Air traffic services are an important feature of the operational environment. They include: air traffic control, weather forecasting and reporting, and aeronautical information systems, with air traffic control being the most important for air safety. As early as 1919, international agreement was reached on the need for standardisation of navigation lights, rules of the air, and signals to be used at airfields. An additional need for ground-based assistance aimed at ensuring safety from air collisions and expediting air navigation, was recognised in the USA in 1936 and in Europe at about the same time.

Air traffic control in the USA

In the USA it was observed that the worst risk of collisions occur near airports, and at intersections of air routes. The first regulation published by the Director of Air Commerce prohibited instrument flight within '25 miles of the centre line of those legs of radio beams regularly used as airways, or within 25 miles of an airline airport'. It was already airline practice to fly on the right-hand side of radio beams and physical features such as railway lines, roads and rivers wherever possible. Three airlines had already agreed on common altimeter setting procedures, all for the purpose of safely separating aircraft.

Air traffic control's first objectives were confined to preventing ground collisions and air collisions during landings and take-offs. Visual signals were used by means of flags, signal lamps and pyrotechnics. Pilots were responsible for avoidance of collisions while en-route from point of departure to destination. Any messages sent from the aircraft by the radio operator were regarded only as progress reports and 'all is well' messages. As aircraft flew in visual conditions most of the time and were few in number and tended to keep separated from each other along the air routes because of inaccurate navigation, there were few air collisions, or even near-collisions.

The USA had the greatest rate of growth in air traffic, and by 1935 there were as many as 50 to 60 landings and take-offs per hour at the Chicago and New York airports. Four airlines, United, American, TWA and Eastern set up a small shared airway traffic control unit in an attempt to regulate the flow of traffic leading into the New York area. In instrument flying conditions, the unit attempted to inform pilots of the progress of other aircraft and the time they would pass over designated points. 'Controllers' were not in direct contact with pilots, and all messages were relayed to and from the aircraft via the airlines' own radio communications systems. It was from this modest beginning that the world's air traffic control system began and, although inefficient, it served its purpose until after World War II, when growth of air traffic and the availability of ground-based radar made improvements necessary and possible.

In 1994, the number of hours flown in the USA were 38.5 million, down from the 44.4 million of 1989. However, there were signs that increased economic activity was leading to resumed growth. Factors increasing pressure on the air traffic control system are a continuing shortage of qualified controllers and the hub and spoke operations of airlines. This leads to pronounced peaks and troughs in movement rates over a twenty-four hour period. Comparisons of system performance are shown in Table 5.1.

Table 5.1 USA Air Traffic Control system performance 1989 and 1994.

	1989 (million)	1994 (million)
Hours flown	44.4	38.5
Pilot reported near mid-air collisions	550	288
Operational errors (less than applicable separation)	912	780
Operational deviations (encroachments of airspace or landing areas)	354	158
Pilot deviations (from FARs)	2478	1291
Runway incursions	223	204
Total system accident data (per 100 000 flight hours)	5.37	5.46

In both the USA and Europe, controversy continues on whether air traffic services should continue to be provided by part of a state authority, or be operated as a separate service. In the United Kingdom, the National Air Traffic Services became a limited company in early 1996 but remained a wholly owned subsidiary of the Civil Aviation Authority. In 1995, annual airport traffic had grown to 900 000 movements at Chicago, 879 000 at Dallas and 733 000 at Los Angeles, with the total for North America being 28.48 million movements. This compares with London Heathrow's 435 000 and Europe's 10.97 million.

An NTSB report in January 1996 showed that 'outages' of major ATC equipment is a major problem. In six months of 1995, ten events totalling almost 195 hours caused 1563 flight delays but only one operational error, largely due to controller skills in reverting to stand-by methods. A major feature of these 'outages' was the age of the IBM 9020E computers and their power supplies.

Steps taken by the FAA to improve system performance are planned expenditures in the National Airspace System Plan ($27 billion) to provide new equipment including primary and secondary radars, and increased computer capacity to aid controllers and new procedures. Research has been carried out to find solutions to problem areas such as wake turbulence, runway incursions, altitude deviations where pilots fail to fly at the correct altitude and near mid-air collisions.

WAKE TURBULENCE. Data shows that between 1983 and 1993 at least 51 accidents and incidents in the USA could be linked to wake turbulence, with 27 persons killed, eight injured and 40 airplanes damaged or destroyed. Boeing conducted tests for the FAA to measure wake turbulence from large airplanes, and a Training Aid was developed for pilots and air traffic controllers. The conclusion reached is that although pressures exist to increase the rate of airport movements, pilots and controllers *must* adhere to the aircraft separation criteria set out in the *Airman's Information Manual*.

RUNWAY INCURSIONS (see Figure 5.1). Airport congestion, non implementation of internationally agreed airport signs, markings, lighting and air traffic control procedures are worldwide problems. Runway incursions have caused a number of fatal accidents, with the worst being at Tenerife in 1977 when 583 lives were lost. In an accident at St Louis in 1994, an MD82 on take-off collided with a Cessna 411, the pilot of which may have been confused between the two parallel runways 30 and 31, one of which he may have mistaken for a taxiway.

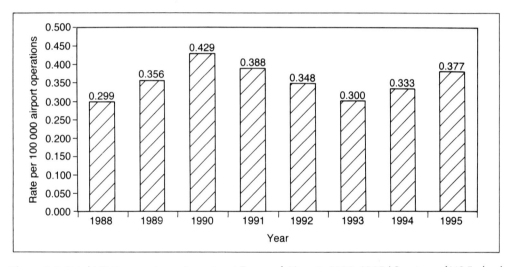

Figure 5.1 Total US Runway incursion rates at Towered Airports 1988–1995 (*Courtesy of US Federal Aviation Administration National Aviation Safety Data Center*).

A 1995 FAA Runway Incursion Action Plan aims to prevent these accidents with improved standards for airport signs and markings; standardised airport procedures; education and communication programs for pilots, controllers and airport operators and a review of movements at 50 high-priority airports. A possible but expensive solution would be to provide surface movement radars to control ground movements at night or in restricted visibility, together with a system that 'tags' returns from airport surface movement radars with aircraft identity. A less costly but valuable initiative would be to provide illuminated stop bars at all runway/taxiway intersections that could be operated by the ATC ground controllers, and to better regulate the timing of complex controller/pilot instructions, so that they are not issued when the pilot's workload is at a maximum e.g. during the landing roll.

ALTITUDE DEVIATIONS. In the 12 years 1983/1994, NASA's Aviation Safety Reporting System (ASRS) received 74 544 reports of altitude deviations, and 87% of these were categorised by ASRS analysts as having flight crew causal factors. Somewhat surprisingly, 60% of altitude deviations were reported to have taken place in later types of airplanes equipped with electronic flight instruments and flight management systems, although these are a minority of the airplane population. Human errors leading to altitude deviations include:

- Controller assigns wrong altitude
- Controller transmits wrong altitude to pilot, who does not read back the assignment and the controller does not request the read back
- Pilot receives correct altitude, reads back the correct altitude but enters an incorrect altitude into the altitude alerter or mode control panel (MCP)
- Autopilot does not capture the MCP altitude setting
- Pilot accepts an altitude clearance intended for another aircraft
- Wrong pressure setting on altimeter, perhaps caused by confusion between millibars and inches of mercury on international flights

Many incidents have multiple sources of errors, and some air carriers have instituted successful altitude awareness programs, including USAIR who achieved a 75% reduction in deviations. When the altitudes occurring most frequently in deviations were examined, 10 and 11 thousand feet were most prominent with 38% of the deviations. These were almost certainly caused by the similarity of one-zero thousand and one-one thousand when passed by radio, and errors in reading three-pointer and counter-pointer type altimeters at these altitudes.

NEAR MID-AIR COLLISIONS. A major factor in reducing reported near mid-air collisions was the introduction of Airborne Collision Avoidance Systems (ACAS), the USA version of which is Traffic Alert and Collision Avoidance Systems (TCAS). Federal Air Regulations require its use by most airplanes in USA airspace, and the version required to be used by scheduled air carriers flying larger airplanes is TCAS 2. This provides separation from all aircraft equipped with appropriate transponders, by providing traffic advisories plus vertical escape maneuvers but with no commands to turn left or right. An improved version TCAS 3, if introduced, would provide traffic advisories plus both vertical and horizontal escape maneuvers, and the USA Air Line Pilots' Association is pressing strongly for its introduction. It is generally agreed that TCAS 2 has proven to be an excellent back-up for ground based air traffic control, helping to detect errors and avoid collisions. However, some air traffic controller organisations remain hostile to the concept, believing that aircraft maneuvering not authorised by controllers may create as many collision problems as it solves.

A less widespread and novel hazard to air safety has arisen from the use of laser displays in cities like Las Vegas. Fifty-one reports were made to the FAA in only two years of pilots being temporarily blinded by intense beams. This led in 1996 to a ban on all outdoor laser displays within a 20 mile radius of airports in Clark County Nevada.

European ATC

Europe has been plagued with fragmentation of its airspace because of national boundaries, differing control philosophies, dominant defense requirements, different languages and computers that could not communicate with each other. When Eurocontrol was created in the 1960s it was neutered by nationalism and confined to control of the upper airspace of three countries, Belgium, West Germany and Luxembourg, with the Netherlands joining in 1987. By this time, Eurocontrol had gained a second role as a research organisation into ATC problems. Political changes including the formation and growth of the European Community into fifteen states, the end of the 'cold war' and rapidly growing air traffic caused a rethink of the role of Eurocontrol. By the late 1980s, severe air congestion caused the 23 member European Civil Aviation Conference (twelve European Union states plus others), to give Eurocontrol the task of replacing national traffic flow management units with a central facility. Other assigned Eurocontrol tasks include:

- Optimisation of air traffic services and airspace structure and preparing the way for the introduction of new technology
- Improving radar surveillance and better sharing of existing radar data
- Reducing radar separation to 5 nautical miles in busy areas
- Progressive integration of ATC systems, to be complete by 1998
- Completion of automatic data communications between ATC centres and Mode S ground/air data link, introduced in the busiest area by 1998.

A report issued in 1992 revealed a formidable task. There were 51 ATC centres using 31 different ATC systems, computers from 18 manufacturers and 22 different operating systems. This resulted in the chairman of Lufthansa complaining that in 1991 his airline lost 13 700 hours flying in holding patterns, wasted 50 000 tons of fuel, and passengers sat in aircraft on the ground for 14 600 hours.

Europe will introduce Global Navigation Satellite Systems (GNSS) for the purpose of improving Control, Navigation and Surveillance, will rely on ILS/MLS for landing systems and will introduce TCAS in the year 2000. United Kingdom commercial air transport risk-bearing near mid-air collision data for 1985–1994 is shown in Table 5.2.

Table 5.2 United Kingdom commercial air transport risk-bearing near mid-air collision (NMAC) data.

	1985	1986	1987	1988	1989	1990	1991	1992	1993	1994
NMACs	16	16	13	20	12	18	6	4	9	15
per 100 000 hours	2.4	2.3	1.8	2.5	1.4	2.0	0.7	0.4	0.9	1.5

ATC in other countries

The quality of ATC around the world is extremely variable, with some developing world countries unable to pay for the facilities and trained personnel required. With most ATC systems manned by civil service employees, pay offered to controllers is often insufficient to retain them. By way of example, a controller trained in computer software programing or similar tasks, is likely to be hired away by banks and financial institutions. Other stress placed on ATC systems is when controllers go on strike and the authority attempts to continue to operate by use of military staff. In 1995, the Venezuelan Defence Ministry took over the airspace and airport control of 20 major airports, causing IFALPA to warn its members that it and ICAO were concerned about safety. In New Zealand in 1995, a strike of controllers resulted in the state introducing an ATC contingency plan that did not have the approval of ICAO, and leading to an unusually blunt exchange of views on whether New Zealand had honoured its international obligations.

In the same year, European pilot associations and Eurocontrol expressed concern about standards of ATC in Greece, with two pilot associations calling the situation 'disastrous'. They cited a survey of pilots, which found that they were frequently unable to understand instructions passed by the Greek controllers. Contributory factors were a long-running dispute between government and controllers, and a shortage of radar cover exacerbated by a dispute between the manufacturer of Athens's single 30 year-old approach radar and the Greek civil aviation authority.

In airspace in parts of Africa, a form of airline-approved 'do-it-yourself' ATC is in use, with pilots using a common VHF frequency (126.9 mcs) to transmit 'blind' their position, height and direction of flight ten minutes before reaching reporting points. Other pilots are expected to note these positions and vary their own flight progress accordingly, all because of inadequate communications and ATC.

At the 1996 annual conference of IFALPA, the whole of Southern Africa was declared 'critically deficient' after the following information was supplied by pilots familiar with operating conditions in the region:

- More than 30 flights per night were unable to make satisfactory contact with air traffic control leading to at least two 'incidents' per night
- Air traffic over Southern Africa increased by 120% in the last five years
- There is little or no radar or VHF communications coverage north of Zimbabwe
- Poor HF radio communications are a contributory factor in numerous near mid-air collisions in recent years, because air traffic control clearances are not provided/received
- Air traffic controllers are unable to contact other controllers in neighboring or distant control centers (even by telephone)
- Training of controllers is inadequate, pay is poor and workload is high
- Many en route facilities are inoperative. (Jeppesen charts are liberally sprinkled with small 'A' symbols showing that a facility has been inoperative for a long period)
- Many airfield facilities are similarly unserviceable
- Airports do not have perimeter fencing and animals and humans wander over operational areas at will
- Radio navigation aids do not transmit identification callsigns
- Aeronautical information services are 'almost non-existent in many parts of Africa'
- Overflight fees paid by operators to states are diverted to non-aviation budget items

Because of this catalog of deficiencies, IFALPA is to hold a regional meeting in Cairo in October 1996 to which ICAO, IATA, states, airlines and other interested parties will be invited.

A newly arrived problem is caused by some states haste in introducing technology such as Global Navigation Satellite System-based Air Traffic Services, without first meeting obligations to obtain ICAO approval, ensure compatibility with other systems in adjoining airspace and the needs of users other than major airlines. A similar controversy surrounds the use of TCAS to permit airplanes to climb or descend through blocks of airspace occupied by other airplanes.

Little progress has been made in many parts of the world toward the ideal of implementing a single coherent ATC system. Hope now rests with GNSS, where air-ground communications and accurate navigation will be assured, and control and surveillance can be remoted from places where power failures and shortages of trained staff are endemic.

ATC in oceanic and other airspace

Almost all aircraft flying over the North Atlantic and Pacific oceans, and through other airspace remote from ground-based navigation facilities, use Inertial Navigation Systems (INS). This provides good accuracy, and would permit reduced separations to be applied if surveillance and direct pilot/controller communications were possible. INS or similarly accurate navigation aids do permit use of Organised Track Systems (OTS), where lateral separation is based on distance, vertical separation is above 28 500 feet at 2000 feet intervals and longitudinal separation is based on time with airplanes required to cruise at fixed mach numbers. These organised tracks are chosen by ATC to take account of prevailing winds and weather, and to take advantage of those ground-based navigation aids that exist and which are published to operators every 12 hours. The system is now (in 1996), inadequate because of increased traffic, but navigational errors are few in number and mostly caused by a pilot wrongly 'copying' his ATC clearance, or wrongly inserting it into the INS or Flight Management System or, more rarely, caused by controller error.

ATC problems of political origin

Some of the more intractable air traffic control problems have political origins, and are not amenable to technical solutions. A long standing example is the one caused by the political problems that exist between Greece and Turkey. The Turkish administration in the eastern part of Cyprus is recognised only by Turkey. It has attempted to gain international recognition by setting up a form of air traffic control system for airspace in the Eastern Mediterranean. However, its authority is not recognised by ICAO, the international airlines or the great majority of pilots who frequently hear, and ignore, 'instructions' broadcast to them by the Turko-Cypriot authorities. These conflict with the information passed to them by the recognised authorities. Other serious problems have existed over the Aegean, where Turkey and Greece disputed the alignment of an internationally used airway, and more recently over Afghanistan and the former Yugoslavia.

Severe problems are posed for civil aviation when hostility exists between states and normal international cooperation breaks down. Contingency plans existing for closure of airspace and absence of normal air traffic control have been brought into use on a number of occasions. For example, when trouble flared up between the state of Israel and its neighbours, and in such circumstances as the warlike hostilities between the states of Iran and Iraq, and Argentina and the UK, etc. On occasion serious disagreements arise between airlines and pilots as to whether operations should continue when the normal standards of air traffic services are not available. In some cases, pilots have refused to carry out normal commercial flights, although they have always agreed to carry out emergency flights for the purposes of relief and the evacuation of civilian personnel.

Aeronautical information services (AIS)

Whenever a pilot departs on a flight, he or she is briefed on the aeronautical facilities available, the forecast weather and also on any other matters of significance for air safety. The collection and distribution of this information is an international task of daunting dimensions, and agreed codes are used to make the task more manageable. Some of the non-routine information is of a nature that requires the use of 'plain language', and states are recommended by ICAO to provide that information in the English language when it is for international distribution. A number of states in South America fail to meet that requirement, and use only the Spanish language causing considerable difficulties for all other nationalities.

Search and rescue (SAR)

The earliest known search and rescue event took place in 1911, when Hugh Robinson, flying in a Curtiss seaplane, rescued a fellow pilot from

Lake Michigan by alighting beside him. The needs of modern aviation obviously cannot be met by relying on the remote and casual possibility of a downed aircraft being seen by the pilot of another aircraft. Therefore, in the event of an aircraft making an emergency landing on water or landing away from an airport, the problem of locating that aircraft as soon as possible is a matter of the highest priority. Some countries with large expanses of undeveloped land require aircraft to be equipped with Emergency Locator Beacons (ELBAs) that are automatically activated in crash conditions. The beacons transmit a radio signal that assists rescue teams to locate the downed aircraft. It is unfortunate that other countries, perhaps with a greater need for such equipment and rescue techniques, do not equip aircraft and rescue teams in a similar manner.

A unique international agreement signed by the USSR, USA, France and Canada in 1979, produced a space satellite solution to the problem of locating activated ELBAs on a worldwide basis, and remarkable success has been achieved with the equipment. Since it began trial operations in 1982, the system has contributed to the rescue of more than 5500 persons in aeronautical, maritime and terrestrial incidents. Six satellites are in operation, and several replacement satellites, incorporating technical enhancements, are being built. Radio frequencies used have been the internationally agreed emergency frequencies of 121.5 and 243 MHz, but it is probable that future satellites and beacons will use the 406 MHz frequency. Maritime organisations participate in the system, and many ships and recreational boats are equipped with the maritime equivalent of ELBAs. The operational advantage of using 406 mcs is enormous because of the lack of congestion in the waveband, with the SAR satellites being primary users, and the reduced possibility of false alerts being generated by the new generation transmitters being used. ICAO, NTSB and the US Coast Guard are in favour, but the FAA is caught in a crossfire between them and the general aviation users who are opposed on cost grounds. The NTSB has stated that there are 400 alerts per day in the USA, and only 10% are genuine. In some parts of the world, civil SAR exists only on paper and those in need of assistance must rely on military organisations or upon the radio watch maintained by merchant ships.

The US Coast Guard operates an Automated Mutual-assistance VEssel Rescue (AMVER) system, that monitors the location of about 2700 ships of more than 124 nations in every ocean of the world. In the event of a maritime rescue becoming necessary the Coast Guard is able to contact the nearest ship and request assistance. As many as 236 000 voyages were logged in 1993, and the system proved its worth on many occasions, being credited with saving 126 lives that year.

SAR is provided for use in many types of emergency, but in civil aviation its greatest value is for the purpose of rescue, and to increase the chances of survival when aircraft 'ditch' in cold water where time is of the essence to ward off the effects of exposure. The term 'ditching' was inherited from the military but survives in civil aviation although it is a euphemism for 'emergency alighting on water' used in the briefings on emergency procedures given by the flight attendants before take-off on a flight that is to include a long over-water crossing. Procedures and equipment used by airlines assume that the ditching may take place in mid-ocean, and aircraft equipment includes: life rafts (or slide-rafts), emergency radios, life jackets, whistles, water-activated lights and emergency supplies of water and food. It is also assumed that sufficient warning of the emergency will be available to enable the flight attendants to brief and prepare the passengers and cabin for the emergency.

There have been a number of emergencies when it seemed likely that a ditching would occur. A B747 that lost power from all four engines due to atmospheric pollution from a volcanic eruption, a Lockheed Tristar aircraft that shut down all three engines while in flight over the ocean but was able to restart one engine and land safely at Miami Airport, a DC 10 that was very short of fuel, and two other wide-bodied aircraft that lost all engine power. These incidents all left their pilots considering the choice of alighting on water rather than on land. Ditchings have been caused by that type of emergency, but more frequently have been unpremeditated, where there was no opportunity to make a choice as to where to land, or to make preparations for the emergency condition. These emergencies usually arise during take-offs and landings at airports situated close to wide rivers, lakes, and shorelines.

In none of these cases did passengers and crew have time to prepare for a ditching, and in some cases lives were lost through drowning. Studies made of these accidents show that the only safeguard possible is to subject the flight and cabin crews to periods of regular training for these types of emergencies. This should be the case even if none of the flights are scheduled over-water as defined in the regulations that determine whether life rafts and life jackets must be carried. It is also clear that airports situated alongside water should include boats or hovercraft in their inventory of emergency equipment.

Airports

Runways, taxiways, aerodrome ground aids and other facilities available at an airport, should match the operational role intended by the airport administration. Should it be planned that operations continue in conditions of restricted visibility, the airport should be provided with Instrument Landing Systems (ILS), together with runway, taxiway and approach lighting to the appropriate standard. Should it be intended to operate in the lower limits of visibility, aerodrome ground movement radar will be a firm requirement if accidents similar to the one at Madrid in December 1983 are to be avoided. In this accident, two aircraft collided on the active runway, with a taxying aircraft inadvertently obstructing the take-off of a second aircraft in foggy conditions.

At airports subject to ice and snow, procedures should be devised and equipment made available to prevent accumulations on the runways and taxiways (see Figure 5.2). Where heavy rainfall is expected, adequate provision should be made to prevent accumulations of standing water by selecting a porous material for the runway surface and by providing adequate drainage. Grooving the runway surface maintains good braking characteristics, and it is surprising that grooving is not invariably provided, because it is a cost-effective means of ensuring safe operations. (The technique is not new, for grooving can be found on Roman roads built more than 2000 years ago.) Frequent removal of deposits of rubber from the runway surface is similarly a cost-effective means of ensuring safe operations.

Acceptable aerodrome standards are generally provided except in countries where funding is not available, or where technical knowledge and administration do not meet requirements. State authorities and airlines are expected to ensure that operations and airports are matched. However, there have been occasions when pilots have felt that they must apply their own pressure in order to ensure that the required standards are met, or that the operations are restricted to those which the airport is capable of safely accepting. Temporary conditions such as power failures are taken into account as they occur, and are common in Africa with stand-by power frequently not being available.

Standards of landing aids at airports directly affect the safety of flight operations, and the Dutch research organisation NLR showed a relationship between standards and accident rates in a 1996 report. It contends that:

- Worldwide, precision approaches (ILS etc.) are 5.2 times safer than non-precision approaches
- Without terminal approach radar at an airport, risk multiplies by a factor of 3.1
- High terrain in the vicinity of an airport increases risk by a factor of 1.2

Figure 5.2 Snow clearance at London (Gatwick) Airport (*Courtesy of British Airports Services*).

- Absence of broadcast airport terminal information systems, or meteorological reports, increases risk by a factor of almost 4
- Absence of published standard terminal arrival procedures increases risk by a factor of 1.6
- Absence of approach lights increases risk by a factor of 1.4

Disparities between the accident rates of ICAO regions are obvious (see Table 5.3), and it is this correlation between aerodrome aids and safety that persuaded IFALPA to 'award' deficiency classifications to airports and airspace, and to use the classifications to apply pressure on the responsible authorities to effect improvements.

Table 5.3 Landing approach accident distribution by ICAO region.

ICAO region	No. of accidents	Movements	Rate/million movements
Africa	17	562 734	30.21
Asia-Pacific	19	1 039 380	18.28
Eastern Europe	5	243 300	20.55
Europe	26	2 732 780	9.51
Latin America	34	1 050 632	32.36
Middle East	3	263 183	11.40
North America	28	6 860 700	4.08
Total	132	12 752 709	10.35

6: Flight Operations

Safety in airline operations depends on the near-perfect functioning of all the many different components of the air transport system. Design, manufacture and maintenance of the aircraft, training and performance of the crew, suitability of the operating environment (including airports and air traffic control) and the operating rules for contending with severe weather are all vitally important parts of the safety equation. Any serious shortcomings of any component will threaten the safety of flight.

The selection and training of flight crew is an important consideration. In normal times there are more well-qualified applicants than there are jobs, but as the airline industry is cyclical in nature, there are occasions when there is a shortage of well-qualified applicants. In some countries the military forces are the back-up for the airlines, and on occasion they are persuaded to release pilots who are willing to join the national airlines. In other countries, pilots normally 'trade-up' their employer, starting in general aviation, moving to a small airline and then having gained the required experience, joining the largest airlines where pay and conditions are better.

Standards demanded of applicants are not fixed in absolute terms, although they must meet minimum licensing standards set by the state. Employers select the best applicants available at the time of recruitment. In some countries major airlines accept *ab initio* applicants and provide all training, sometimes on a pay-back basis and always with a contractual obligation to serve the employer for a fixed period after gaining qualifications. Two-pilot crews create a need for increased training before a 'new-hire' becomes operational, as there is no option of on-the-job and under-supervision learning. The use of sophisticated flight simulators in pilot flight training has therefore become common, and many hundreds are in use worldwide. A typical large airline such as United, had 26 full flight simulators in use in early 1996.

Crew complement

Post World War II airlines operated four-engined aircraft carrying thirty or forty passengers over long distance worldwide routes. It was common practice to operate with three pilots, two flight engineers, one navigator and one radio operator. By the middle 1980s, airlines operated four-engined Boeing 747–400 series aircraft capable of carrying in excess of 500 passengers with a two-pilot flight crew.

The process of increasing the size and performance of airliners while simultaneously reducing the number of crew members, took almost forty years and involved an unusual mixture of technical and socio/industrial developments. The first specialist crew member to be made technologically redundant was the radio operator who left the airline scene when high-frequency radio-telephony replaced radio telegraphy during the period of the late 1940s and early 1950s. The navigator departed approximately ten years later when authorities, airlines and airline pilots agreed that developments in radio navigation techniques and the installation of ground-based long range navigation aids had removed the need for a full-time navigator. The last routes to need navigators were flights in high latitudes, where reliance could not be placed on magnetic compasses and where the work-load involved in navigation consumed all the time of one person. The adoption and use of Doppler and Inertial Navigation Systems (INS) and procedures, capable of precise navigation on a worldwide basis was the final nail in the coffin of specialist navigators. It has to be said that air navigation became much more precise when sophisticated machines replaced men in the navigation task!

By the 1960s, two-engined aircraft were normally operated by a two-pilot crew. The rationale for this crew complement was that two-engined aircraft had fewer systems and controls, and were operated over shorter and better organised route systems than three and four engined aircraft. They were also never distant from alternate airfields, and were therefore able to land without delay in the event of any abnormality arising during a flight.

It was in the 1980s that the aircraft manufacturers began to design and manufacture large two-engined aircraft capable of flying long distance routes. These routes were formerly the preserve of three and four-engined aircraft, and a serious dispute arose between pilots and flight engineers on the one side and the aircraft certification authorities, airlines and aircraft manufacturers on the other. Organisations representing pilots and flight engineers made public statements claiming that a two-man crew complement was less safe than a three-man crew, and produced arguments and scenarios to support their case. In a number of airlines, industrial action was taken by the flight crews in support of their position, and in some countries a three-man crew for the new aircraft was conceded by the airlines.

In the USA, the dispute became so rancorous and public that a Presidential Task Force was formed in March 1981 to determine whether a new version of the two engined DC9 (the MD–80 series), could be flown safely by a two-man crew. The national pilot association, ALPA, agreed in advance that it would accept the recommendations of the Task Force as being final. The Task Force hearings were of great importance and interest to the worldwide civil air transport industry, and a number of foreign airlines, aircraft manufacturers and organisations representing flight crews took part or submitted documentary evidence.

When the Task Force issued its report in July 1981, it was clear that the evidence had been carefully considered. Although the report stated that the MD–80 could be safely operated by a two-man crew, it recommended that steps be taken to meet some of the concerns raised by the pilots. It was specifically recommended that the FAA procedure for determining the minimum crew safe for the operation of an aircraft should be improved and strengthened in several respects.

These events created new methodology for determining crew complement and led to important changes in pilot training, airplane systems, controls and displays, with increased attention being given to human factors in airplane design. That the changes were successful cannot be contested, for the air safety record has improved in recent years. However, pilots have reason to believe that their early opposition to two-pilot crews was a critical factor in ensuring that the change was carefully implemented.

Controlled flight into terrain (CFIT)

In 1931 a tri-motor Fokker crashed into Australia's Snowy mountains, and since that date it is estimated that more than 30 000 passengers and crew have lost their lives in terrain-related accidents. In the last 20 years, more than half of the fatalities in western-built jet airliners have been attributable to CFIT in which serviceable airplanes have hit the ground despite being under the control of their crews. Because of this appalling record, special measures were needed to remove this hazard. In the early 1970s Scandinavian Airlines System (SAS) originated the concept of a Ground Proximity Warning System (GPWS), that would alert the pilot of an impending flight into terrain. The system uses a computer to process radio altitude, barometric altitude, rate of descent, deviations from ILS glide path, aircraft configuration (position of flaps and landing gear) and air data computer information, to generate warnings. In late 1974, 90 people died when a B727 struck the top of a ridge only 20 miles from its destination – Dulles Airport, Washington. This accident prompted the FAA to enact regulations requiring large jet and turbo-prop airplanes to have GPWS by the end of 1975. In the same year the British CAA mandated GPWS installations in all large commercial jet airplanes, and in 1979 ICAO implemented internationally agreed GPWS standards, thus attacking the CFIT problem on a worldwide basis.

The rate of CFIT accidents in the USA fell from 0.6 per million departures in 1975 to 0.1 in 1995, a factor of six, during a period when flight sectors doubled. This reduced accidents by a factor of 12, an obviously remarkable achievement. Because regional airliners and business jets had a CFIT record 40 times worse than airliners, and lost an average of three airplanes per year, the FAA now requires all turbine-powered airplanes with more than 30 passengers to have GPWS installed.

Early GPWS suffered false alarms that adversely affected pilot confidence in the systems, and in some instances airplanes crashed while pilots attempted to establish whether an alarm was genuine. A series of improvements to the computer software reduced this problem, and pilots are now expected to respond immediately to GPWS warnings, by applying full power and effecting an immediate maximum rate of climb.

A Dutch study of CFIT accidents, reported by the FSF, has stated:

- 40% of all approach/landing CFIT accidents occur in areas where there are no significant terrain features near the airfield
- Aircraft operating in the African region of ICAO, face the highest risk of 0.7 CFIT accidents per million departures
- 75% of 108 aircraft for which equipment data was available, were not fitted with a GPWS
- Of the 27 aircraft that were fitted, most were of early Mark 1 & 2 types of equipment and of these no warnings were generated in 33% of events.

Two efforts, one led by ICAO/IFALPA and a second by the FSF, were in 1993 combined into a joint effort that sought the help of the manufacturers (mainly Boeing) to attack the CFIT problem. The pilot training aids and simulator training programs resulting from this combined effort are in use by most airplane operators today. The next development of GPWS will be to add information from a digital terrain database together with positional information from the airplane navigation systems, to provide up to 40 seconds (compared with 10 seconds) warning of high ground ahead. In early 1996 United Air Lines and British Airways had each installed a trial system in one airplane. It may eventually be possible to depict the terrain immediately surrounding the airplane on a Head Up Display (HUD) so making every flight the equivalent of a Visual Flight Rules (VFR) flight, as regards CFIT.

Flight simulators in pilot training

Greatly improved flight simulators (see Figure 6.1) made it possible for the USA's FAA to authorise their use for almost all aspects of flight training. This includes:

- A pilot's initial conversion to a different type of aircraft
- Periodic checking of competency
- Instrument rating qualification
- Upgrading to Captain status
- Renewal of 'recency' qualification
- Training in emergency and windshear procedures
- Training for 'all weather' operations in reduced visibility

This remarkable step was achieved because of the success of airlines and simulator manufacturers in working together to build the flight simulators that almost perfectly reproduce the appearance, flying qualities and sounds of aircraft that occur in typical airline flight operations. The simulators also satisfactorily create other stimuli, such as visual and movement cues, that are needed by the pilot if flight simulator training is to be near-identical with that carried out in flight. Simulators do not yet play a significant role in *ab initio* pilot training, which continues to use simple types of aircraft. Yet in mid 1987 Swissair became the first airline to use full-flight simulators for *ab initio* training.

By the late 1980s the FAA initiative was clearly a success. Many states followed its example, with airlines gaining an improved standard of training at reduced costs, with B747–400 flight simulator time costing only 5% of aircraft flight time. A universally welcomed side benefit arising from an increased use of flight simulators is a dramatic reduction in the number of serious aircraft accidents resulting from flight training, particularly when simulating engine failure on take-off. During the period 1958–1980 there were 65 jet transport training accidents on a worldwide basis, and 31 of these accidents occurred during training for engine failures, with 27 aircraft being destroyed.

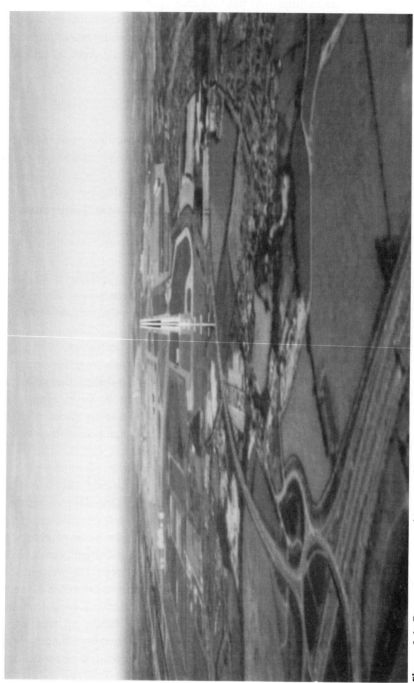

Figure 6.1 Computer generated image of London Airport (*Courtesy of Thomson Training & Simulation*).

Benefits obtained from the use of flight simulators include: the avoidance of air traffic congestion and noise nuisance at training airports, improved pilot training (particularly in those rarely encountered emergencies that are too difficult or dangerous for in-flight training), increased safety and greatly reduced costs. In the 1990s, an international standard for flight simulators was agreed, thereby simplifying the design, sale, approval and use for multinational pilot training.

Training for in-flight emergencies

Emergencies too dangerous to perform in flight are demonstrated on simulators until the required standard of pilot proficiency is reached. This feature is of great value in training to contend with such hazards as: engine failure on take off; emergency descents due to loss of pressurisation, fire or smoke drills; and more recently, encounters with severe windshear.

A training exercise that may be scheduled, or perhaps be requested by a pilot, could include the total incapacitation of a pilot, followed by smoke or fire from electrical sources on the flight deck. This would leave the remaining pilot to assume a huge workload associated with a statistically remote possible emergency that creates difficult problems.

Pilot incapacitation is simulated by briefing one of the pilots to imitate subtle incapacitation, perhaps during an approach to land, when the workload on both pilots is at its highest level. The role-playing pilot usually quietly ceases to take any part in the operation of the aircraft and does not give any indication of the problem to the other pilot. Although cases of pilot incapacitation are rare, they are believed to have been a causal factor in some accidents, and training for such eventualities makes a significant contribution to air safety.

A high priority task when a pilot becomes incapacitated in flight is to have flight attendants remove him or her from the seat, or to secure him or her by use of the full safety harness. Depending on the circumstances, the remaining pilot will have already declared a full emergency to ensure that the air traffic service units provide maximum assistance so as to reduce cockpit workload. Should a further emergency such as cockpit fire and smoke from electrical sources occur, the pilot immediately dons a full face mask and goggles, breathes pure oxygen and declares an emergency by transmitting the international distress call, *Mayday Mayday*. The pilot then informs ATC (and the cabin crew and passengers) of his or her intentions, and initiates an emergency descent if the fire and smoke emergency condition does not end when the emergency drill is performed.

Wearing a full face mask and goggles causes difficulty in communicating with ATC, cabin crews and passengers, and the flight deck loudspeaker system is therefore used to speak to the flight attendants. Smoke would make the task of performing emergency drills, reading flight instruments and making control selections, extremely difficult. Hence British Airways simulates the visual characteristics of smoke by adding layers of cling-film to the pilot's goggles, making his or her vision extremely poor.

In addition to the task of flying the aircraft, the pilot: selects maximum cockpit lighting, has the cockpit/cabin door closed, isolates the defective electrical system (shedding load if necessary), switches off the recirculating fans, reduces cabin pressure by selecting an increased cabin altitude and a maximum rate of cabin climb, carries out an emergency descent (after checking sector safety altitude) and selects a suitable and nearby airport for an emergency landing. The workload incurred is enormous and a pilot is only able to cope by making optimum use of the resources available to him or her.

Modern airplane systems reduce pilot workload in these circumstances. The automatic flight control system will perform almost any task provided the appropriate selections are made, with the navigation system displaying all emergency airfields suitable for that type of aircraft, together with the navigation aids and radio frequencies required. Should smoke persist in the flight deck area or passenger cabin, a pilot would almost certainly opt to carry out an automatic approach and landing. He or she would then have the flight attendants prepare the passengers for emergency evacuation of the aircraft immediately it came to a stop on the runway.

Simulation of the scenario described here is extremely realistic, except that it is not possible to produce smoke in the simulator. Performing the training exercise stretches a pilot to the maximum and leaves him or her with many items on which to reflect. The pilot would probably end his or her period of reflection by hoping that such an improbable set of adverse circumstances never happens, and certainly not before he or she has amassed considerable flight time on the aircraft type. An ability to operate all the controls from memory and touch, without being able to see, would enormously improve chances of surviving such an incident!

Windshear training

Flight simulator windshear exercises programed with data taken from accident reports severely test pilot skill, with the relationship between take-off thrust and weight making the exercise more difficult to fly on three or four engine airplanes than on two-engined airplanes. In encounters with severe windshear in both take-off and landing cases, pilots are taught to extract maximum performance from the airplane and engines and to fly just above stall warning speed until out of the danger zone. Better forecasting of the phenomenon and widespread use of air-borne windshear prediction systems are needed, if these accidents are to be eradicated.

Great effort is made by all airlines to impress on pilots that windshear is a potential killer, and that the first objective of a pilot must be to *avoid, avoid, avoid*. The incident that happened to a B727 aircraft at Denver is frequently quoted as an event that would almost certainly cause any aircraft to crash. It is estimated that the aircraft encountered a 60-knot tailwind over the last 7000 feet of the runway during take-off, causing it to collide with the ILS antennae and only narrowly avoid disaster. Another extreme case quoted is the Tristar accident at Dallas in August 1985. Here the core of the downburst that caused the fatal accident was estimated to be descending at 3000 feet per minute. This resulted in the aircraft having a final rate of descent well beyond the performance reserves of any type of airliner.

Pilots are warned that until the forecasting and detection of this phenomena is improved, it is likely that they will experience an encounter at some time during their career. They must learn to recog-nise the onset of an event and be able to fly out of the encounter when it happens. It is emphasised that a typical landing approach is performed at near-standard pitch attitude, airspeed, rate of descent, and power setting, and that whenever a significant excursion from those conditions occurs, windshear should be suspected. Should the windshear occur at an early stage of the approach it may be possible to fly safely through the event, because by definition, a microburst has a diameter of only two nautical miles. Pilots are instructed that if windshear is encountered during the later stages of an approach they must 'go around', i.e. carry out a missed approach. Pilots are reminded that during take-off, pitch attitude to achieve a safe rate of climb varies only with gross weight, air temperature and airfield elevation, and again any significant excursion from the norm should be taken as being indicative of possible wind-shear.

It is probable that after having reviewed a windshear training period, many pilots will reflect that windshear has always existed although it was not recognised as such until a few years ago, and that some earlier accidents assigned to 'pilot error' should have had windshear recorded as a prime causal factor.

Emergencies in flight operations

Training is provided for flight and cabin crews to enable them to safely operate aircraft in a hostile environment, and to save lives on those occasions when a real life emergency happens. Similar considerations apply to the provision of aircraft and ground safety equipment and procedures.

In-flight fires

Over a period of 25 years, airlines suffered approximately 50 cabin fires in flight of which about half were fatal causing about 1000 deaths. The three worst happened at Paris in July 1973 to a Boeing 707, at Riyadh in 1980 to a Lockheed Tristar and at Cincinatti in 1983 to a DC 9. All aboard the 707 died, 301 died in the Tristar and 23 in the DC 9, a total of nearly 450 people. In all three cases, the aircraft made safe landings in difficult circumstances but lives were lost from the heat, smoke and poisonous gases emitted from the burning cabin furnishings. Examination of the circumstances show that there are lessons to be learned, and improvements in equipment and procedures to be made if such accidents are not to happen again in the future. It is surprising that some airlines continue to use passenger blankets made of polyester that is easily ignited. The following are items that are receiving urgent attention.

Fire protection equipment

Fire extinguishers carried on board airliners must be suitable for fighting fires of electrical and other origins. One of the most effective fire extinguishants is water/glycol but this is not suitable for fighting fires of an electrical origin, and so Halon extinguishers are provided for fighting fires on board aircraft. Halon 1211 is extremely effective but lethally toxic in concentrations greater than 5% for periods longer than five minutes. That level of concentration could easily be reached in confined areas such as cockpits, galleys and toilets, and for that reason some airlines have turned to the use of Halon 1301 which is safe at concentrations of 7% for as long as 30 minutes. Halon extinguishants are 'unfriendly' to the environment and are to be phased out over a period of several years.

Regulatory authorities require smoke detectors to be fitted in aircraft toilets. Flight attendants make frequent checks of these compartments in flight, as it is known that the fires in the B707 and DC 9 started in the toilet compartments at the rear of these aircraft. At the same time improvements are being sought in furnishings used in airline aircraft. The FAA have introduced rules that require 'fire blocking' materials to be incorporated into seat cover materials. Other items of cabin furnishings must also be demonstrated resistant to flames and heat sources so as to increase the amount of time available for aircraft to land and for the passengers to be safely evacuated. Experience has shown that normal cabin lighting, even when powered from emergency sources, is almost always invisible to passengers and crew members when the aircraft is filling with smoke at head level, and purpose-designed escape lighting at or near floor-level is required.

Smoke hoods

The best means of enabling a person to breathe for a period long enough to escape from a smoke-filled aircraft cabin, would be for passengers to use personal and portable oxygen systems similar to those available for some crew members and invalid passengers. However, the complexity, weight and cost of such systems make smokehoods a better choice. Oxygen sufficient to survive three to five minutes is present in human lungs, and FAA researchers suggested using that oxygen source by adding a readily donnable simple bag, non inflammable and heat reflective, with a see-through visor and flexible seal to close the bag around the neck to trap the uncontaminated rebreathable air inside.

During the period 1988–91 regulatory authorities in the USA and Europe expressed interest in passenger smokehoods and published draft specifications that manufacturers were able to meet, but after representations from airlines it was decided that smokehoods would not be required because 'they would slow down cabin evacuations'. The UK Civil Aviation Authority claimed that a study over a 20 year period showed that they might save one life but possibly result in eight deaths in an average year. Some frequent travellers disagree and buy smoke hoods for their own use!

Factors in favour of smoke hoods include, delaying the onset of panic induced by the hot and toxic irritant smoke which affects both breathing and vision, extending the survival time in a toxic environment by several minutes and retaining passengers in a conscious and mobile state. Possible disadvantages are: inducing a false sense of protection in passengers perhaps slowing down the rate of evacuation, condensation inside the masks reducing vision, and the attenuation of communications.

Effective protective breathing equipment is carried on the flight deck of passenger airliners for the flight crew to fight in-flight fires. In mid 1987 the FAA extended that requirement to include similar equipment to be carried in the cabins of aircraft to enable the cabin crew to also fight in-flight fires. The portable breathing equipment is to be located within three feet of each fire extinguisher required to be carried by earlier FAA regulations. Although the British CAA has not made the provision of breathing equipment for cabin crew mandatory, British Airways adopted corporate policy to provide breathing equipment for *every* crew member of *all* aircraft types operated, and believes that availability of the equipment will ensure faster and safer emergency evacuations of its aircraft in a fire or smoke contaminated incident.

The more difficult problem of preventing smoke and noxious gases from reaching the cockpit in flight, has yet to be solved and the report of the DC 9 pilot in the Cincinatti accident shows that the aircraft was almost lost due to the appalling concentration of smoke on the flight deck. In 1991, the FAA approved the use of an Emergency Vision Assurance System (EVAS), that deploys an inflatable vision unit between the pilot's goggles and vital flight instruments within 15 seconds of being activated ensuring continued viewing of the instruments. Inflation is by filtered air free of particulates and aerosols. The manufacturer points out that there are approximately 600 occurrences per year of in-flight fires, smoke, fumes and explosions, with more than 100 occurring on the flight deck with 34% of them resulting in emergency landings, but the system was not in widespread use in 1996.

Aircraft sprinkler systems

There have been many successful demonstrations of effective cabin sprinkler systems (see Figure 6.2) that use normal aircraft domestic water supplies to feed a specially designed nozzle system that delivers water sparingly in droplets of the right size and speed to cool an aircraft cabin fire and absorb its gases. This then provides sufficient time for aircraft passengers to escape in an emergency and protect them from lethal doses of smoke and fumes. The spray rate is only 13 gallons per minute through the special nozzles, which are fed by three supply pipes along the cabin ceiling. Airfield fire and rescue teams can plug their water hoses into the system and ensure that the drenching flow continues after the aircraft water supply is expended. In tests observed by civil authorities, airlines and pilot organisations, extremely intense aircraft cabin fires have been quenched in only three seconds.

Arguments against installation and use center upon water damage to electrical systems (any accidental discharges requiring 100 man hours and more than two days of down time), weight increases and costs. The FAA, and UK and Canadian regulatory authorities had draft regula-

Figure 6.2 Demonstration of the sprinkler system showing severe fire, eleven 'passengers' on board (*Courtesy of Save Ltd*).

tions ready in 1993 but eventually agreed that improved flammability requirements for cabin fittings removed the need to proceed, although they were agreed on the effectiveness of the systems. The wisdom of that decision will be tested by future statistics of deaths from airplane fires (see Figure 6.3).

Rescue and fire fighting

Great efforts and expenditures are made by all components of the civil air transport industry in support of accident prevention programs, but it is inevitable that accidents will continue to occur in such a complex activity that is conducted in a hostile environment. In recognition of that inevitability, aircraft are designed for crashworthiness, and a percentage of airport resources are provided for the sole purpose of rescue and fire fighting. Special equipment is carried on aircraft for use in emergencies, but notwithstanding all these efforts, human lives continue to be lost in circumstances that are foreseen and are in some cases avoidable. Some of the problem areas that remain are at aerodromes, some are in aircraft and some are related to the training and equipment made available to flight crews and personnel of the emergency services and to the relevant safety briefings provided for passengers.

In 1981, a study of cabin safety in large airline aircraft in the USA showed that 77 survivable/partially survivable accidents/incidents had

Figure 6.3 Severe fire damage but a survivable environment maintained in the passenger cabin (*Courtesy of Save Ltd*).

occurred since 1970. Table 6.1 shows the location of those accidents by proximity to airports.

The data shows that a high standard of fire fighting and rescue resources at airports is essential to the saving of lives, and so tables of recommended emergency equipment and resources for different categories of airports are published by ICAO. Unfortunately for air safety, each state may vary these recommendations and the actual provision made varies widely. In some countries major airports have little or no dedicated fire and rescue equipment and rely on equipment and personnel that may be positioned at a town that is miles away. As the chances of surviving a post-crash fire is inversely proportional to the time taken to evacuate the aircraft, it is clear that a crash and fire that occurs to a large aircraft at one of these airports will almost certainly result in heavy loss of life.

ICAO Standards require that aerodromes establish an emergency plan, procedures be adopted to test the plan (by conducting a full-scale emergency exercise at least every two years) and a review of the results be

Table 6.1 Location of accidents by proximity to airports.

On airport	1–5 miles	Greater than 5 miles	Total
45	12	20	77
58.44%	15.58%	25.97%	

made. Guidance is provided on how to meet these standards and it is recommended that suitable rescue equipment and services be provided at aerodromes located close to water or swampy areas where a significant number of approach or departure operations take place over these areas. It is permissible to use off-airport facilities if a response time of not more than three minutes to the end of each runway and to other parts of the movement area can be achieved. Speed of response is held to be essential if rescue and fire fighting is to be successful and a maximum two minute response time is the objective. Recommended scales of equipment reflecting the category of the aerodrome are set out by ICAO.

The reason that all these provisions are necessary may be shown by considering the following two examples of many accidents. In Los Angeles in 1978 when a DC 10 ran off the runway and caught fire, an achieved 30 second response time saw all passengers (except two senior citizens who fell off a wing into a fire area), safely evacuated. In 1993 a B747 overran the runway at Hong Kong, fell into the bay and a 300 person rescue boat was available. There have been other similar accidents at waterside airports including La Guardia and Kennedy, and in such accidents amphibious rescue vehicles would be of great value.

Even where a state adopts the ICAO recommendations, they are based on the use of a table that takes into account the number of 'movements' (take-offs and landings) and the overall length and maximum fuselage-width of the largest airplanes using the airport, to determine the level of protection to be provided. The formula then permits the use of a remission factor that lowers the protection to be provided by one category, when the number of movements of the largest aircraft is fewer than 700 in the busiest three consecutive months of operations. This remission factor reduces the costs to the airport operator but means that an accident to a very large aircraft at that airport would have levels of emergency equipment appropriate for a smaller type of aircraft. Naturally this would be of little comfort to the passengers on board a large aircraft in an emergency situation, to find that because only 699 large aircraft used the airport in the three busiest months, their chances of survival had been consciously reduced to a lower level!

In the USA and Western Europe public airports are licensed as a means of achieving and maintaining acceptable levels of safety, and the factors considered include the provision of emergency services. Emergency exercises are conducted to test the equipment and procedures and any deficiencies found must be rectified. The provision of safe overrun areas and the removal of dangerous obstructions in the runway area receive particular attention, and features found at some foreign airports would not be accepted.

When operations at an airport are conducted in conditions of very poor visibility, fire and rescue vehicles should be equipped with the means to locate an aircraft in a distress condition. At some airports use is made of the airport's ground radar plus a radio link, in order to provide the required navigational ability in conditions of restricted visibility. Other authorities have experimented with the use of Forward Looking Infra Red (FLIR), to help rescue vehicles locate crash sites and navigate around airports in poor visibilities. On occasions helicopters have been of great value in emergencies, with down-wash from the rotors being very useful to fire fighters.

Aircraft cabin equipment

An USA NTSB study published in 1981 showed:

- Components of seat/restraint systems failed in 84.4% of the aircraft accidents examined
- 77.7% of overhead panels, racks and passenger-service units failed in the accidents and caused injuries to the passengers or 'hampered emergency egress'
- Components of galley equipment failed on 62% of occasions and sometimes blocked emergency exits.

In an accident to an MD80 in Sweden all overhead bins became detached together with attached passenger service units including oxygen units, reading lights and air vents although the 'g' forces experienced were probably less than the 6 g static load requirement. It was suggested that there is need for dynamic testing of these items and the NTSB expressed concern about the same problem affecting other types of airplane (see Figure 6.4). The 1981 NTSB study was similarly concerned about seat failures on impact.

Since the 1981 report was issued, much attention has been given to these problems and the FAA required that improved seats be fitted to airplanes built after 1988, after calculating that they would have saved 107 lives and 63 serious injuries over an earlier 14 year period. At about the same time the British CAA proposed that cabin floors be strengthened, aisles widened, exits improved and in some cases a seat removed from near an exit after investigating accidents where high g loads were experienced or evacuation rates were shown to be in need of improvement. A more recent rule seeks to avoid head injuries to passengers. No civil authority or airline has yet adopted rear facing seats, although expert opinion is almost unanimous that they are safer, and the USAF

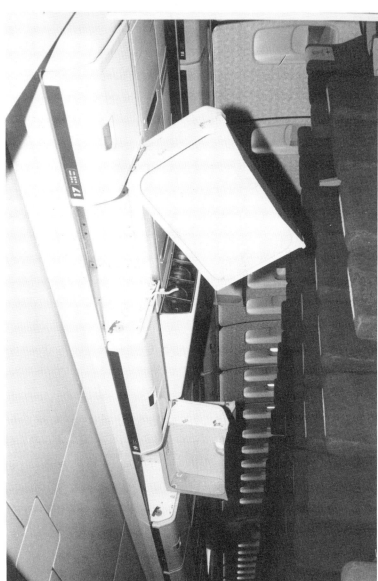

Figure 6.4 B747 at San Francisco. Cabin fittings failed under high *g* in flight (*Courtesy of NTSB*).

insists on their use in military transport aircraft. New research into crash impact forces show that a 'legs back' brace position is best for forward facing passengers, optimum seat pitch is 32 inches and three-point seat belts would reduce injuries.

A serious omission is the provision for child safety although regulations exist for motor vehicles and it is certain that children in restraint systems are much safer than when held by adults. Air accident statistics show that children are particularly vulnerable in airplane accidents. In December 1978 in a DC8 crash in Oregon, three of the ten fatalities were children under two years, and in an accident to a DC9 in 1987 with three children and two lap-held infants on board, one child and one infant received fatal injuries. The only action taken by regulatory authorities and airlines to date is to permit passengers to use their own carry-on child restraint systems but not to make their use mandatory. Child restraint systems available to the public vary widely in effectiveness.

Safety problems can arise with the carriage of passengers with disabilities, and American Airlines has expressed concern that in some cases aviation safety is being compromised. Examples provided are wheelchairs with liquid batteries, use of supplemental oxygen and the effect on evacuation rates of the carriage of mobility impaired persons. It is standard practice in US airlines that physically-handicapped persons are not seated in rows adjacent to normal and emergency exits as they may not be able to operate exit openings. (In some countries it is required that a cabin crew member be seated near to *every* exit.) As there are 47 million people in the USA with disabilities of various kinds, it will be seen that these problems are real but with no immediate solutions in sight.

Aircraft emergency equipment

Large passenger aircraft carry such items of emergency equipment as: fire axes, extinguishers suitable for fighting electrical and other types of fires, emergency oxygen supplies, loud hailers for use in crash landings or ditchings, life rafts, flotation vests, escape slides (sometimes combined with the life rafts), medical kits, and on some routes, emergency supplies of food, water and survival clothing. Normal items of aircraft equipment such as radios and exterior lights also play an important role in an emergency when locating the aircraft is a matter of high priority and most flight and cabin crews receive some training in use of the equipment in emergency conditions.

The FAA has a rule requiring that medical kits be carried on all large passenger aircraft for diagnosing and treating passengers who may suffer heart attacks. Yet the prevalence of litigation and claims for damages in the USA courts when negligence or malpractice may be charged has resulted in a marked reluctance of medically qualified persons to provide their services in emergencies. In only two years, the FAA recorded 2322 medical emergencies on US domestic flights resulting in 33 deaths.

A survey made in 1982 showed that heart attacks accounted for 28% of serious incidents, nervous attacks (fits etc.) for 15% and some 12% were gastro-intestinal problems. Other common causes were listed as psychiatric, respiratory, obstetric and gynaecological. One airline has an average of between six and ten in-flight deaths per year, and another reported that a doctor was required to treat a passenger on 166 of 11 000 long haul flights.

In Europe, regulations will come into force in 1997 requiring that should a suitably qualified person be on board an airplane and a passenger or crew member needs medical attention beyond simple first aid, some additional medical equipment should be available to enable that person to capitalise on his or her skills and perhaps save life. The Captain will have authority to release a cockpit-stored emergency medical kit that contains more sophisticated equipment and drugs than does a first aid pack. The medical kit will contain 23 items but some airlines already exceed that requirement, and British Airways carries 88 items plus a baby-delivery pack, perhaps because in one year its crews dealt with 2 105 incidents among 30 million passengers.

In one well publicised case, 88 items were not enough and a surgeon-passenger operated on a female passenger suffering from a collapsed lung, using scissors and local anaesthetic from the kit, a wire coat hanger and a bottle of brandy. The coat hanger was straightened and used to push a plastic tube into the patient's chest cavity to inflate the lung, brandy was used to sterilise scissors and the operation was successful. Some long haul airlines have doctors on call on the ground 24 hours per day, contactable worldwide, and crews are given first-aid training and refresher courses. A medical association magazine has stated that an average of 750 people die from in-flight heart attacks each year but only two airlines carry defibrillators.

Emergency exits and evacuation procedures

Numbers, types, size and positioning of emergency exits are determined by regulatory authorities and in recent years there has been agreement between the FAA and Europe's JAAs on requirements for new types of airplane, but differences continue as to 'grandfather rights' for types with long production lives due to the steady improvement of a basic design. The B737 is an example of this, with a 1960s FAR (that was withdrawn about fifteen years later) allowing an additional five pas-sengers per exit pair when approved automatic evacuation slides were incorporated for these exits, allowing ten more passengers per plane than allowed for by the basic exit arrangement. This continued to be allowed on all current (1996) 737 variants with the same exits, and was initially claimed to apply to the next generation of B737s, the 600, 700 and 800s although never claimed for B757/767s or Airbus designs. A particular irritant to Europe's JAAs is that the increased number of seats is rarely used in the USA but is widely used by European charter-flight operators. A welcome compromise was agreed in 1996 to apply to the B737 600/700/800 airplanes and ensure similar treatment for these types and narrow-body Airbus airplanes.

Although passenger evacuation trials are conducted during certifica-tion testing of airplanes, and full evacuation of passengers and crew in 90 seconds with 50% of the exits blocked must be demonstrated, tests cannot be truly representative of real emergencies. During research trials performed in the UK, financial rewards were offered to the first 50% of *passengers* to exit the airplane and this introduced an element of com-petitive behaviour and stress that, on three occasions, lead to *passengers* becoming jammed in the exits and the test being called off. This beha-viour did however enable performance comparisons to be made between different cabin layouts, aisle widths, seat pitch and seat placement at exits, and these and other tests led to improved European regulations to deal with these items with FAA rules meeting the same objectives.

Analysis of accidents and research programs has shown that the role adopted by cabin crews greatly affect evacuations, with 'assertive' crews speeding up the process. Research jointly sponsored by the FAA and the UK Civil Aviation Authority showed that using a single exit, the average time to evacuate the first 75% of passengers was 79.68 seconds with no cabin crew, 76.59 seconds with two non-assertive crew, 63.2 seconds with one assertive crew and only 57.97 seconds with two assertive crew. These results justify an increased use of cabin simulators (some with a moving base and able to introduce smoke) used for the training of airline crews in evacuations. A Canadian study of 21 emergency evacuations showed that 'in at least four occurrences, the absence of effective crew commu-nications may have placed both passengers and crew in positions of unnecessary risk', pointing up a need for emergency procedures training for complete crews.

Computer modelling of cabins and exits is a promising technique. It enables cabin layouts to be changed easily for simulator tests of proce-dures and rules made to contend with occasions when an exit is inop-erable and a flight continues with a reduced number of passengers. The number of passengers that may be safely carried is determined, and the seating plan modified to provide the safest possible arrangement.

In mid-1996 the FAA was researching the 'fire proofing' of airliner hulls using a new material, a heat-stabilised oxidised polyacrylonitrile fibre with a high carbon content that may delay burn-through for 5 minutes, thus doubling the time available for passenger evacuation and fire fighting in an external fire situation. The new material is said to be lighter than the glass fiber-reinforced plastic batting normally used for the insulation of hulls.

7: Costs Versus Air Safety

It is widely believed that regulation, economic strength and safety are interrelated and that it is impossible to achieve an acceptable level of air safety without taking them all into account. In May 1969, a Committee of Inquiry into British Air Transport under the chairmanship of Sir Ronald Edwards KBE, stated on page 1 of its report, Cmnd. 4018:

'The need for governmental regulation of air safety standards has been accepted for many years. Indeed the unanimity of governments on this point is now such that in many aspects of air safety it has been possible for governments to agree internationally, under the aegis of the International Civil Aviation Organisation the standards which they will enforce. Long and patient negotiation has produced these results, which represent a real achievement of international cooperation. With the continuing and rapid growth of international air transport, the maintenance of international air safety has become a matter of great importance from which no government could lightly abdicate...

'The ways in which the enforcement of air safety standards in this country, including the adoption of internationally recommended standards, may be expected to influence the financial situation of the airlines, which will be the subject of study later in this Report. At the outset, however, we shall say very briefly that the complete devolution of responsibility for deciding and enforcing safety standards onto the individual airlines or to the airlines collectively would not be acceptable.

'The view is very widely held that market forces alone could not be expected to elicit from all airlines at all times a sufficiently high and consistent degree of attention to air safety standards. It is true that in the long run a good accident record would serve an airline well commercially, but it would be unthinkable to most people that government should abandon the principle of prevention based on regulation and certification and wait for an airline to disqualify itself through its accident record from the confidence of its customers.'

On page 210 of the Report is a paragraph headed:

'Safety Performance as a Function of Economic Strength.'

'There is a general belief, shared by most of the many persons and organisations in various parts of the world with whom we have discussed the subject, that there is a relationship between the "attitude" of airlines to operational standards and their economic strength.

'Economic weaknesses might have an effect on safety in two ways. First, there are those cases where an airline suffers from under-capitalisation or insufficient profitability. These could restrict investment in items whose relevance to the actual operation is not directly obvious but which are necessary to create the right operational environment to match the scale of activities to which the airline is committed: hangarage, workshop and office facilities, training schools and the development of a ground organisation to provide adequate back-up for the operation.

'Secondly, there could be a strong temptation, in the short-term at least, to cut back expenditure on crew training and engineering maintenance because immediate savings can be made in these areas without affecting revenue. Staff at all levels will sense the economic situation and some, perhaps against their better judgement, may feel the need to adopt procedures which benefit the company's financial situation in the short-term but which also jeopardise the safety position. We are thinking of the aircraft commander who feels himself under pressure to land in marginal weather conditions and so avoid the heavy expense involved in a diversion to an alternative airfield, or the engineering staff who might decide to put into service an aircraft whose airworthiness condition was questionable.

'The arguments against these possibilities are, first, the Flight Operations Inspectors should detect them, and, secondly, that everyone in the industry is concerned to ensure the highest standards of safety, none more so than the airline operators and staffs themselves, who have so much at stake that they would be the last to take chances.

'It is extremely difficult to find hard evidence from British experience that there is a correlation between financial strength and safety. This is only natural for an operator who, for example, cuts corners in operational or engineering activities is hardly likely to issue a staff directive on the matter. Moreover, these situations probably take time to develop; the individual steps in the regression are difficult to discern and perhaps of no great significance in isolation. Where management is less than competent, there may even be a lack of awareness that the economies may have an effect on safety standards. There is also an understandable reticence in the disclosure of information on this sensitive subject, since some of it could be construed as defamatory.'

Under the same heading the Report states:

'It is significant too that on five occasions when the Director of Aviation Safety was about to withdraw an Air Operator's Certificate this was found not to be necessary because the operator was going out of business for financial reasons. However, we have to admit that our conclusions on this subject are largely a matter of judgement or "feel". But it is judgement based on a broad concensus of opinion especially of authorities responsible for the enforcement of safety standards in a number of countries...'

In early 1987 United Air Lines questioned the US Department of Transportation's right to require financial reports from airlines for the FAA's use in allocating safety inspection resources. The DoT responded: 'FAA has found a relationship between a carrier's financial position and its safety record', thereby justifying this requirement.

Deregulation in the USA

Deregulation was introduced into the USA by the Airline Deregulation Act of 1978 and in the 1990s it is possible for almost any organisation to enter the USA civil air transport industry in much the same way as they could enter the trucking industry, and to gain rights to operate domestic air routes that used to be protected from undue competition. The remaining major constraint is the allocation of 'gates' at busy airports, and that process is subject to purchase by bidding at especially arranged auctions or by sale from one airline to another.

Air passengers enjoy lower fares that are available on some routes as a result of fare wars between airlines, and in order to make low fares possible and stay in business, airlines seek major reductions in costs by imposing lower salaries and wages on a reduced number of employees. These changes have caused industrial unrest and a number of older airlines suffered damaging strikes called by unions that were unwilling to see the high standards of pay and conditions formerly enjoyed in the airline industry disappear without a fight. Newer airlines tend to be non-union and these differing labour policies cause difficulty when a non-union and a union airline are merged.

Deregulation-caused airline mergers and takeovers continue into the 1990s and may end with only three or four mega-carriers surviving. Pioneer airlines such as Eastern and Pan-American disappeared in controversial circumstances, with Eastern suffering fines for safety violations and a long and fatal strike by employees. Safety violations uncovered during this period caused scrutiny to be made of the FAA safety oversight program, and in 1989 the General Accounting Office (GAO) published two reports that found FAA aviation inspectors not receiving adequate training and its safety inspection management system lacking adequate oversight. After these reports were published the FAA introduced changes and regained its reputation, although arguments continue as to whether it should become an independent organisation and lose one of its roles of safety regulator *and* protector of the commercial wellbeing of the aircraft manufacturing and operating industries.

Other consequences of deregulation were development of hub and spoke route operating patterns and takeovers of regional airlines to provide traffic feed to large airlines at those hubs. The practice of 'code sharing' where an airline sells tickets that involve travel on another related carrier and the passenger is not necessarily informed that two or more carriers and/or types of airplane are involved in his/her journey with varying standards of flight operations, is common practice.

Deregulation and safety

USA ALPA, the largest member association of IFALPA, was greatly affected by deregulation of the USA's civil air transport system and in 1984 stated its views in the following terms:

'The proponents of deregulation continually point out that legislation only addresses the economic regulation of the airlines: the safety regulations will not be changed. We do not disagree with this assertion, but we do take issue with the conclusion that safety, therefore, will not be affected. To take comfort in the regulations and enforcement responsibility of governments is to avoid the safety question...

'The primary factor in airline safety is cost. Maintenance of high safety standards is extremely expensive, whether measured in terms of equipment, facilities, maintenance reliability or by operating procedures and conditions. This fact is often taken for granted because it is rarely visible to the consumer of airline services. Nonetheless, airline companies have invested millions of dollars in on-board safety systems, personnel training facilities and programs, and aircraft maintenance and overhaul bases, which for the most part cannot be translated into economic efficiency. It is simply the price which must be paid to achieve the highest possible degree of safety in airline operations.'

Discussing new-entry airlines ALPA stated:

'they will not bring with them into the business the enormous investments in safety facilities and programs, and experienced personnel which have been a major part of the overhead of established airlines for many years...

'Extraordinary vigilance by government should be mandatory in a deregulated environment to guard against a relaxation of safety practices and procedures, particularly in that segment of the industry which lacks experience in air transportation and a long standing commitment to safety.'

Support for the gloomy expectations of organised pilot groups on the safety consequences of deregulation came from other organisations dedicated to the cause of air safety. The Flight Safety Foundation (FSF) at its annual meeting in October 1985 listened to a great deal of supporting testimony. The President of the FSF Mr J. Enders said, 'The architects of deregulation failed to apply the fault-tree analysis technique which the aircraft designer is expected to apply to an aeroplane', with the result being, 'sharp economic competition which cannot enhance safety'. Commenting on staffing changes resulting from economic reorganisation, Mr Enders said they would 'provide opportunity for operational error'.

A 1984 Labor Department survey showed that USA airlines cut more than 4300 mechanics at line stations between September 1980 and June 1984. Although airlines added mechanics at maintenance depots, the additions were too few to make up the difference. Overall, the survey shows the number of mechanics employed by the airlines decreased by at least 3000. A trend to decrease spending on maintenance coincided with a period of increased airline revenues and as a percentage of total operating expenses, maintenance decreased from 8.85% to 7.6% of the budgets of US airlines during the early 1980s.

Gerald M. Bruggink, formerly of the NTSB, wrote in a 1991 FSF publication:

'there is a distinct change for the worse in the accident experience of US air carriers in passenger operations during the second half of the 1980s. The contribution of deregulation to that record could be substantiated only in those accidents where flight crew experience was identified as a factor. This does not mean that experience levels in other areas, maintenance for example, have not been affected because it is mainly the flight crews' experience that gets routine scrutiny in accident investigations.'

Mr Bruggink went on:

'the Airline Deregulation Act of 1978 was a well intentioned but draconian experiment implemented with some naive assumptions about the infrastucture's capacity to absorb the anticipated growth in the traffic and the FAA's ability to monitor the burgeoning industry. By giving free rein to the driving forces of economics and corporate ambition, the industry entered a period of upheaval whose full impact on the competence and the commitment to safety of the work force may not be felt until the 1990s.'

Mr Bruggink also expressed the opinion:

'Before deregulation it would have been unthinkable that crew experience would be a factor in Part 121 (*major airline*) accidents. Carriers voluntarily went beyond the prescribed minimum qualification standards. However these practices may have been compromised when the explosive increase in airline operations during the mid and late 1980s required the massive hiring of ground and flying personnel.' (He then described two accidents supporting his opinion.)

The FAA became increasingly active in its role of guardian of air safety, and in mid 1990 carried out an audit of airplane production line assembly procedures that uncovered evidence of 'unsatisfactory' quality control leading to Douglas and Boeing improving some of their practices. In late 1992 Delta Air Lines was fined $1.5 million for maintenance 'discrepancies' and it was obvious that this safety oversight activity was necessary and could be effective.

By November 1993, Mr Bruggink was able to write in an FSF publication that in the period 1990–1992, USA airlines had achieved their best three-year performance in air safety with fatal accidents, measured against departures and hours flown, the lowest ever, although financial losses reached an all-time high. Mr Bruggink stated that fatal accidents in 1990–1992 showed an absence of documented deregulation and proven maintenance factors.

Other causes for concern in the early 1990s were that the drive for lower costs had resulted in USA commuter and regional airlines being substituted for major airlines on short sector flights and to act as 'feed' to major airlines at hubs. These flights were increasingly operated on a code sharing basis with major carriers, and to passengers there was little to distinguish commuter operators from patron airlines, except by their inferior safety record. From 1990–1992 commuter airlines were more than nine times more dangerous than major airlines in terms of hours flown per fatal accident, and more than four times more dangerous in terms of fatal accidents per departures.

This was partially attributable to rapid expansion and operating under less demanding FARs than major carriers. This discrepancy in safety regulation attracted the attention of US ALPA who initiated a 'One Level of Safety' campaign that won support from the FAA and resulted, in 1996, in commuter airlines being brought under most of the regulations applied to major carriers, FARs Part 121. A second safety problem was a high attrition rate of regional pilots which in 1990 reached 50% per year in some companies, an impossible burden for managers responsible for pilot recruitment and training. American Airline's solution to this problem was to integrate pilot recruitment and training programs of its regional airlines and AA's so that recruitment is to the regional airline with progression to the major at a later stage so providing an incentive for pilots to join and remain with the regional airline.

The May 1996 crash of a ValuJet DC9 in Florida had early repercussions when the FAA revealed that it had forced the airline to reduce its planned rate of growth several months before the crash, and had expressed concern about the airline's contracted-out maintenance arrangements. The airline suspended operations and a political and safety debate about FAA's twin roles of promoting the cause of aviation in the USA *and* to regulate safety was reopened. Mr Broderick, the internationally respected Associate Administrator responsible for regulation and certification resigned his post, and many observers agreed that he had become a 'scapegoat' for an administratively impossible situation and that his abilities would be sorely missed.

Another manifestation of safety problems may arise where there is 'code sharing' between airlines of two or more countries with very different safety cultures in order to achieve lower costs. So-called globalisation is widespead with alliances between USA and foreign carriers and again a passenger may be unaware when buying a ticket that travel is on carriers and airplanes of which he or she has no knowledge. It is possible that in future more operations will be switched to the partner carrier with lower costs, perhaps because that partner experiences fewer safety inspections and oversight. It is important for safety that civil air transport does not become like the maritime industry where 'flags of convenience' result in huge fleets of ships registered in countries where regulation does not exist.

The ultimate form of deregulation occurred with the collapse of the Soviet Union, where a monolith state and airline splintered into many states and hundreds of airlines operating in primitive operational environments with virtually no safety oversight, resulting in a poor safety record. There have been persistent reports that control of some of these airlines passed to criminal organisations that threaten and coerce employees, including pilots, to ignore such regulations as may exist with airplanes being grossly overloaded and flight crews kept ignorant of what is being carried. Availability of aviation fuel is not assured and one report suggests that an airliner experienced a CFIT accident because it was so short of fuel, the crew had to reduce route mileage near the destination airport. Other reports tell of more passengers being carried than seats provided. Somewhat surprisingly, the United Nations has chartered some of these 'airlines' to fly its missions to places of famine or civil unrest and the safety record of the flights fall short of normal civil aviation standards. (See Table 8.6 in Chapter 8.)

The operational environment was affected when deregulation caused a large increase in flights in busy airspace and to hub airports. Since USA domestic airlines were deregulated in 1978, followed shortly by some liberalisation in Europe, competitive policies in aviation have been adopted by governments throughout the world. Competition between privatised and national carriers has been fierce and heightened by economic recessions, with emerging states continuing to place a high priority on air transport development but with limited funds to support it. These combined pressures result in increased traffic but reduced safety margins, with a tendency to reduce operating criteria to minimum permitted standards:

- At airports, approach separation distances are reduced
- Pilots are under pressure to reduce the time they occupy runways by using high speed turnoffs and taxy at higher speeds
- Separation distances between taxiways and runways are reduced to provide more space on the parking apron
- Traffic growth requires more controllers to be trained and the average level of experience is reduced
- Movements per hour per runway are increased by sequencing

arriving airplanes at minimum separation distances, allowing airplanes to land immediately after each other or after airplanes taking off, increasing the risk of exposure to wake turbulence

■ Separation minima between parallel runways is reduced and they continue to be used in instrument flight conditions with increased risk of collision

■ Operations from intersecting runways are allowed, increasing the risk of collision

■ Increased air traffic increases radio communications on already congested channels and causes misunderstandings between pilots and controllers

Solutions to some of these problems are to globalise air traffic management systems, improve the standard of flight operations safety management, have pilots and controllers adhere to regulations and make a greater investment in applied technology to reduce the possibility of human error causing accidents.

'Affordable' safety

Lessons learned in the USA as a consequence of deregulation were noted by the international air transport industry, but may have been ignored by states which look with favor on the political philosophy of less governmental involvement in what they see as 'just another business'. The USA tends to set standards for the rest of the world because of the huge size of its aircraft manufacturing and operating industries. Many states, other than those in Western Europe, regulate air safety by simply adopting the USA's Federal Air Regulations, without unfortunately possessing the same levels of technical competence and experience to administer and interpret them. International statistics show most of the aircraft accidents that occur in the world, except for the Soviet Union, cannot but portray the very different attitudes and abilities that exist in airlines and administrations.

When Concorde was designed, the achievement of maximum payload was realised to be much more critical to the success of the project than with subsonic aircraft. A table of costs-benefits was produced (see Table 7.1).

The USA nominal value of a human life is $650 000 in 1983 dollars, and is used when calculating cost-benefits of proposals to introduce new safety equipment or procedures. The Department of Transportation requires that no new regulation shall be allowed unless the savings

Table 7.1 Cost-benefit assessment, Concorde safety equipment.

Safety equipment	Lives likely to be saved per $ million investment
Escape slides	7.0
Oxygen	1.5
Life jackets	7.0
Life rafts	0.2
Cabin fire extinguishers	1.5

(Ground Proximity Warning Systems (GPWS) are estimated to save nine lives per $ million on the same scale.)

exceed the costs. Using this criteria, the Director General of Airworthiness of the UK CAA produced the following example of a calculation for the purposes of deciding whether to proceed with a proposal to require that fire-blocking material be used in the manufacture of seats for passenger aircraft to be used in the USA.

- Additional time available for passenger evacuation   43 seconds
- Lives saved   (13.6 at $650 000 each) $8 840 000
- Damage prevented $2 210 000
- Annualised total $11 050 000

Similar calculations were performed on the cost of adopting the use of anti-misting kerosene as fuel for aircraft, and the costs came out as high as $750 million over the first ten years in the UK alone. The Director General stated that in his opinion that amount of money could be spent more usefully on other programs to improve air safety, as the saving of life might be restricted to only one in twenty accidents.

A regulation adopted in the UK improved requirements for passenger access to emergency exits, and the total costs to operators came out at approximately £2.5 m. That cost was seen as being justified by experience gained in the fatal accident to a Boeing 737 aircraft at Manchester Airport in August 1985, when difficulties were experienced in opening and removing some of the exits.

Human errors cost the airlines a great deal of money, and sometimes cost passengers their lives. In 1983, a Tristar operated by a USA airline had a triple engine failure because O-ring seals were omitted from the oil systems when minor servicing work had been performed on all three engines, and it was only good fortune that prevented a major disaster. When the USA's NTSB enquired of the FAA some months later, what had been done to prevent a re-occurrence of the error, it was informed that the Engineering Vice Presidents of the airlines that operate that type of engine had all been informed. The appropriate rejoinder made was that Vice Presidents do not carry out servicing work! A very different response of a European airline was to modify the part involved so as to positively prevent it from being replaced without the O-ring seals, at a cost for the whole fleet of £125 000.

A cost-effective tool for safety is the Maintenance Recorder, or Airborne Integrated Data System, sometimes called AIDS. The system has the ability to monitor engines and safety critical systems in a way that cannot be achieved by human observers. It detects trends and gradual deteriorations of performance, and when a pre-determined level has been breached, alerts the engineering branch of the airline. The capital investment required can be great but large airlines use them to good effect and are gaining benefits in safety and reduced engineering costs.

The costs of airlines are only partially under their direct control, and the industry has always suffered from low profit margins. In the new and more competitive climate engendered by economic circumstances, some different and new problems have arisen with a need to reduce operating costs in order to remain competitive and to meet changed circumstances. Basic cost items such as aviation fuel vary in price according to political factors as well as supply and demand, and the operations departments of airlines respond to higher prices by reducing the amounts of reserve fuel carried on aircraft.

Aircraft hull and liability insurance premiums paid by airlines have greatly increased and those that insure against sabotage, hijacking and war risks are finding it difficult to meet the costs. When, following an accident or other event causing loss of life, litigation takes place in the USA, awards made by the courts have been staggering. In one case of an accident to a general aviation aircraft, the plaintiffs were awarded $4 m in compensatory damages and $25 m in punitive damages, for one life. In addition to increasing the costs of insurance, such awards severely limit the flow of safety information, for manufacturers are unlikely to wish to publicise defects in their products if the consequences include litigation and excessive awards in court cases.

In mid-1996, FAA attention was focused on the operations of ValuJet, a low fare airline, after one of its airplanes was destroyed in an accident over the Everglades in Florida. The Wall Street Journal, in its edition of 6 June 1996, stated that the FAA criticised ValuJet for its practice of paying pilots only for completed flight segments. This unusual practice results in pilots not being paid for cancelled operations due to mechanical problems, and it was alleged that one flight continued to Nashville with a gear problem instead of returning to Atlanta. By mid-June it was reported that ValuJet had suspended operations.

8: The Accident Record

The construction and operation of aircraft made rapid progress under the stimuli of World War I, and even before post-war reconstruction was properly begun the first airline flights were made. Several airlines vie for the honour of being first, but it is generally agreed that Avianca of Colombia and KLM of the Netherlands were among the first. The earliest British scheduled flight took place on 25 August 1919 when a DH4a aircraft left London (Hounslow) for Paris (Le Bourget), carrying one pilot, one press representative, newspapers, leather, several brace of grouse and a number of jars of Devonshire cream.

Since that early beginning many airline aircraft have been destroyed by accident, lives have been lost, and many causes have been listed for the accidents, including the following:

- Structural failure of the aircraft (overload, corrosion, fatigue, stress, vibration)
- Engine failure
- Failures of flight control systems
- Loss of control by the pilot (disorientation, failures of flight instruments)
- Bird strikes
- Fire
- Weather conditions (fog, wind, turbulence, snow and ice, lightning strikes)
- Fuel (shortage or contamination)
- Human error by the flight or ground crews
- Air collisions caused by pilot error and/or errors made by air traffic controllers
- Ground collisions at airports
- Sabotage.

The list is incomplete and many accidents are attributed to combinations of several causal factors. Studies made of accidents enable the design of aircraft and the training of personnel to be improved, but as one problem is solved another appears because of the ever changing operational objectives and environment, design objectives, and materials used. Supersonic passenger aircraft operate at the speed of a rifle bullet at the highest levels of the atmosphere, and structural materials have progressed from canvas covered timber through alloys, stainless steel, titanium and on to reinforced plastics, each producing new problems of design, manufacture, maintenance and repair.

A significant development since the advent of jet powered aircraft in the late 1950s, is that because a performance plateau for airline aircraft has been reached, they are expected to use up *all* of their safe lives although earlier aircraft were retired after only a few years of operation due to becoming obsolete. Airliners are now required to operate at high rates of utilisation for periods of 20 years or more. Additional factors likely to adversely affect safety are the serious financial problems of airlines being coincident with a tendency for states to relax their previously firm regulation of the airline operating industry. These factors taken together may cause accidents at a rate that reach the boundaries of acceptability.

ICAO tables for worldwide scheduled services during the period 1985–1995 show a generally improving trend with acceptable deviations occurring in some years because of the statistically small number of accidents (see Table 8.1).

Statistics for 1995 indicate that it was a better year than 1994 for both worldwide and USA air safety (see Figures 8.1, 8.2 and 8.3).

Safety in 1995

The ICAO annual report for 1995 providing information on aircraft accidents involving passenger fatalities in scheduled services, showed that safety levels are significantly different for various types of aircraft. In turbo-jet aircraft operations which account for 95% of total passenger kilometers, there were 11 accidents with 541 passenger fatalities. In turbo-prop and piston-engined aircraft operations which account for 5% of traffic volume there were 17 accidents with 169 passenger fatalities. The fatality rate for propellor-driven airplanes is much higher than for turbo-jets, but the type of power plant used is only one factor in producing a discrepancy, and type of operation flown and the operational environment may be equally significant.

Table 8.1 Fatal accidents, worldwide scheduled services, excluding CIS.

Year	Aircraft accidents	Passengers killed	Fatalities per 100 m passenger km	Fatal accidents per 100 m km flown	Fatal accidents per 100 000 aircraft	
					hours	landings
1985	22	1066	0.09	0.21	0.13	0.19
1986	17	331	0.03	0.15	0.09	0.14
1987	24	890	0.06	0.20	0.12	0.18
1988	25	699	0.05	0.19	0.12	0.18
1989	27	817	0.05	0.20	0.12	0.19
1990	22	440	0.03	0.15	0.09	0.15
1991	25†	510	0.03	0.18	0.11	0.18
1992	25	990	0.06	N/A	0.11	0.16
1993	31	883	0.05	N/A	0.19	0.21
1994	28	941	0.05	0.14	0.14	0.15
1995	26	710	0.03	0.12	N/A	0.13

† includes one collision on ground – shown here as one accident.

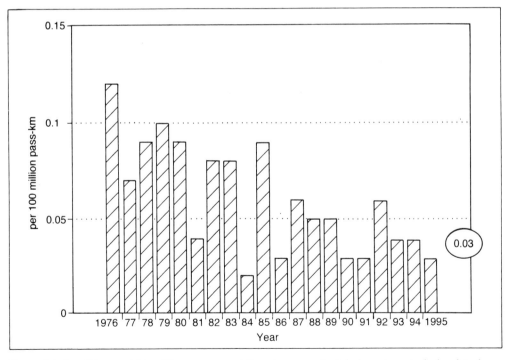

Figure 8.1 Fatalities per 100 million passenger kilometers on scheduled services. Includes data from the CIS from 1986 onwards. (*Courtesy of ICAO.*)

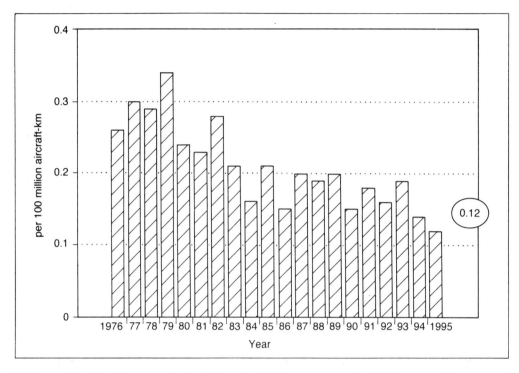

Figure 8.2 Fatal accidents per 100 million aircraft kilometers on scheduled services (*Courtesy of ICAO*).

Types of accidents

During the period 1988–1993, worldwide airline fatalities classified by type of accident (see Table 8.2) showed that 76 accidents cost 3513 lives

The table shows why the ICAO/FSF program to avoid CFIT accidents is so important.

European JAAs made an analysis of causal factors in 219 fatal accidents recorded by ICAO and the UK Civil Aviation Authority, and showed 824 causes, an average of about 3.8 per accident, as shown in Table 8.3.

It is significant that 'failure to crosscheck/coordinate' is judged to have been a factor in 118 out of 219 accidents (54%), and this was related to 'lack of situational awareness' the next most frequent item, suggesting that maximum use of all navigational means is of prime importance. The analysis suggests that human factors, flight deck skills, professionalism, CRM and training are the keys to reducing the number of accidents and it is more important to focus on how airplanes are operated than on the airplane itself.

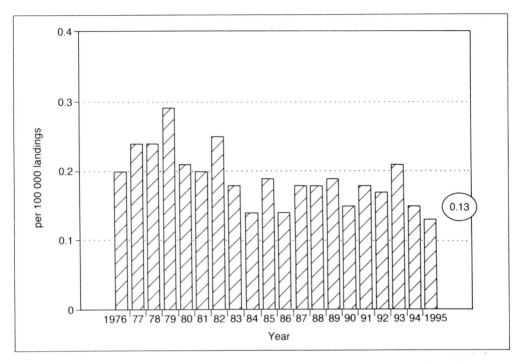

Figure 8.3 Fatal accidents per 100 000 landings on scheduled services (*Courtesy of ICAO*).

Many of these items fall within the responsibilities of management, and increased attention is given to the thesis that safety must be accepted as a responsibility of top management before it can be delegated to operatives such as pilots, engineers and air traffic controllers. In 1993, the President of the FSF stated that 'the 70, or so, percent of fatal accident causes (primary or probable) ascribed to pilot or flight crew

Table 8.2 Classification of fatal accidents 1988–1993.

Cause	No. of accidents	Fatalities
CFIT (controlled flight into terrain)	28	1883
Loss of control (airplane caused)	10	460
Loss of control (crew)	14	357
Airframe	4	278
Midair collision	1	157
Ice/snow	4	134
Fuel exhaustion	5	107
Loss of control (weather)	2	79
Runway incursion	3	43
Other	5	15

Table 8.3 Ranking of causal factors.

Factor (group)	No. of occurrences	% of accidents
Failure to crosscheck/coordinate (crew)	118	54
Lack of situational awareness (crew)	75	34
Flight handling (crew)	60	27
Omissions/wrong action (crew)	57	26
Collision with high ground/obstacles	49	22
Current safety equipment not fitted (aircraft systems) GPWS/TCAS etc.	47	28
Failings leading to collision with level ground/airport	35	16
Non fitment of new equipment (enhanced GPWS)	31	14
Failings leading to impact with obstacle/ obstruction	28	13
Lack of ground aids (ATC/ground aids)	24	11
Post crash fire	24	11
Slow/delayed action (crew)	23	11
Poor visibility	22	10
Engine failure	22	10

error is now being unmasked as a misleading statistic'. He went on to state that management inattention is the greatest aircraft accident cause, not pilots. An earlier President of the FSF suggested many years ago that the Chief Executive Officers (CEOs) of the airline, manufacturer, FAA or any other significant party be called as first witness in any NTSB public hearing concerning a major accident.

The FAA is to develop regulations 'in 1996' that will require all commercial carriers to have independent 'airline (flight) safety departments' and when the FAA made that announcement, members of the Air Transport Association and Regional Airlines Association stated that they are to name officials who will report to the CEO to provide a top management focus on safety.

A 1993 study (see Table 8.4), by McDonnell Douglas of commercial jet transport aircraft accident statistics over a period 1958–1993 analysed according to aircraft systems, engine related accidents and control problem accidents produced interesting data.

Landing gear accidents dominate this table and should concern safety regulators and manufacturers!

When worldwide hull loss accidents for commercial jet fleets are analysed for the period 1959–1994 per flight segment (see Table 8.5), the following is a result.

Table 8.4 Analysis of 35 years of commercial jet transport accidents 1958–1993.

Aircraft systems		Control problems		Engine related	
Landing gear	456	Pilot induced	124	Uncontained failure	63
Engine	192	Weather induced	72	Engine fire/warning	47
Fuselage	185	Other system involved	41	Power loss	43
Wings	109	General	35	FOD†	39
Flight controls	89	Aircraft structure	25	Inflight shutdown	37
Hydraulic power	49	Uncommanded action	25	Flameout	31
Structures	42	Aircraft stall	22	General	31
Equipment/furnishings	40	Autopilot failure	10	Engine departed	22
Power plant	39	Weight/CG problem	10	Engine stall/surge	21
Nacelles/pylons	37	Locked controls	8	Multiple engine failure	16
Fuel systems	33	Buffet/vibration	7	Thrust reverser	15
Doors	27	Wake turbulence	1	Engine fuel problem	5
Stabilizer	22			Throttle/autothrottle	4
Engine exhaust	21			Oil system	3
Navigation	16				
Electric power	15				
Autoflight	13				
Air conditioning	13				
Windows	10				

(† FOD = Foreign object damage.)
Each event may involve more than one system, therefore the items may be more than the total accidents of that type.

International comparisons

Disparities exist between the safety records of airlines of different countries, and because human ability, natural environment and the airplanes used are much the same, the disparities most probably arise from differing operational environments, limited training resources, absence of safety oversight and an inadequate investment in safety

Table 8.5 Worldwide commercial jet fleet – accidents v flight segment.

Segment	% of accidents	Exposure, % of flight time
Load, taxi, unload	1.9	0
Take off	14.2	1
Initial climb	10.1	1
Climb (flaps up)	6.7	14
Cruise	4.5	57
Descent	6.9	11
Initial approach	11.4	12
Final approach	24.3	3
Landing	20.1	1

mechanisms. National sensitivities being what they are, international organisations are reluctant to address these problems. A 1984 study of the previous ten years showed fatalities per million flights varying from 0.656 for Australia (the best) to 509.693 for Colombia (the worst). If continents were looked at, the following table of merit resulted:

- 1) Australia
- 2) North America
- 3) Europe
- 4) Asia
- 5) Africa
- 6) South America

but there were big variations between countries within those continents. Significant factors affecting comparisons for the period 1984–94 are the collapse of the USSR and an unprecedented rate of growth of aviation in China, both accompanied by increased accident rates for which full statistics are not available. Unofficial data for accidents involving aircraft manufactured in, or operated by, states of the former Soviet Union (including helicopters) is shown in Table 8.6:

Table 8.6 Accidents involving aircraft manufactured in or operated by former states of the USSR.

Year	Accidents	Fatalities
1992	92	536
1993	45	579
1994	34	565
1995	20	280
1996†	6	465

† first five months

This accident rate is poor over the whole period but shows a strong improving trend.

In the April/June 1987 edition of the 'Boeing Airliner' magazine, results of a survey of 12 'better than average' operators of Boeing airplanes were examined. Over a ten year period, 16% of all operators of Boeing airplanes had crew-caused accident records higher than the average of all operators of the airplanes and accounted for 80% of these accidents. If it were possible to gain access to hull insurance rates quoted to airlines, big differences would show.

It is these facts that caused the USA to examine the safety oversight programs of countries that have airlines operating to the USA and to deny or reduce access to the USA market to those failing to fulfill a criteria of meeting ICAO standards. European countries and ICAO have taken similar steps aimed at eventually improving the worst safety records (see Chapter 3).

An IFALPA publication of March 1996 reported a presentation by Captain A.H. Faizi, Chief Pilot, Corporate Safety of Pakistan International Airlines. Highlights of Captain Faizi's presentation were the following:

- In the past 12 years commercial airlines had 222 hull losses
- 50% of the world's commercial fleet is owned by 20 airlines and they suffered 34 hull losses
- The remaining 50% is owned by 152 airlines and they suffered 188 hull losses

Captain Faizi concluded that safety management is an area of weakness in 'developing world' airlines and that common problems include:

- Lack of enforcement of regulatory functions
- Inadequate professional training
- Non-professional safety management
- Scarcity and/or mismanagement of funds
- An ageing fleet
- Operator and regulatory body placed under one bureaucratic head, who is chief executive of the airline and also heads the regulatory authority
- Military men appointed to run airlines
- Absence of independent regulatory bodies

Captain Faizi quoted the International Airline Passenger Association's (IAPA) 1995 grading of safe/unsafe airlines (see Table 8.7), showing India and Colombia as 'two of the most dangerous places in the world to fly', with each of those countries having a fatal accident rate ten times higher than the rest of the world and 20 times worse than airlines on its list of 16 outstanding airlines.

None of these airlines fly a 'developing world' flag and neither do any of the next 32 airlines on IAPA's listing of airlines.

Air safety comparisons: USA versus the rest of the world

In June 1995 the FSF published its own examination of safety statistics produced by Boeing, and commented that the worldwide rate of fatal

Table 8.7 IAPA Safety Listing of Airlines (1995).

Major 2 m flights annually	Mid size 1–1.2 m flights annually	Med–Small 1 m flights annually
American	All Nippon	Alaska
British Airways	American West	Finnair
Delta	Ansett Australia	KLM
Lufthansa	Canadian Airlines	Malaysian
SAS	Saudia	Swissair
South West		

accidents in 1994 remained within a range that has varied little since the 1970s. A comparison of fatal accidents for USA and non-USA operators of commercial jets, showed a long-term pattern in which the rate for non-USA operators exceed that of USA operators but with the trend lines paralleling one another fairly closely, especially in the early years of commercial jets (see Figure 8.4).

Dangerous goods and air safety

Major causal factors in aircraft accidents and incidents include: human factors, airworthiness of aircraft, acts of terrorism, failures of the air traffic services and natural hazards such as severe weather. Yet other factors are also important to those who wish to maintain safety in airline operations. Safety is threatened by many substances carried aboard aircraft in the form of: passengers' baggage (whether in the hold or cabin), legitimate air cargo, airmail letters and parcels, aircraft stores and equipment, explosives placed on board by saboteurs or hijackers and even by the very large quantities of fuel needed for flight.

An adverse trend arises from a great increase in the number of potentially dangerous materials produced for use by all manner of legitimate businesses. Although the list of dangerous substances created by the chemical industry is revised on a frequent basis, it is difficult for the most diligent shippers and carriers to keep up-to-date on materials and the safest methods of handling and storage. For many years IATA exercised responsibility for determining the standards of packaging and the restrictions on carriage applied, in order to maintain air safety. Its list of Restricted Articles was accepted in most parts of the world and used by almost all shippers and cargo-carrying airlines whether members of IATA or not.

In 1983, ICAO published regulations on Dangerous Goods and effectively took over from IATA. The reasons given by ICAO for taking

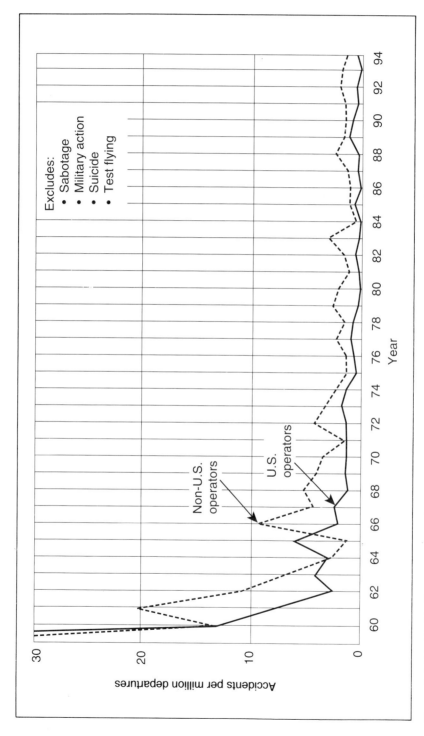

Figure 8.4 Worldwide Commercial Jet Fleet Accidents, US and non-US (*Courtesy of Boeing Commercial Airplane Group*).

responsibility, were the need to harmonise transport by air with other modes of transport, to achieve compatibility with other international regulations published by the United Nations and the International Atomic Energy Authority, and to give the regulations a higher legal status.

The use of different terms in civil air transport for what are basically the same materials, is undesirable. The term 'Dangerous Goods' used by ICAO will probably prevail over IATA's Restricted Articles in the long term. A third term, 'Hazardous Materials', is used in a more general sense. It is accepted as the term to identify substances or materials that are determined to be capable of posing an unreasonable risk to health, safety and property when transported in commerce, by any mode of transport, and which have been so designated. They are defined as: Explosives, Combustible Liquids, Corrosive Material, Flammable Liquid, Flammable Gas, Organic Peroxide, Oxidising Materials, Poisons, Irritating Materials, Etiologic Agents, Radioactive Materials and Other Restricted Materials (ORM).

Familiar items are also included in that definition: matches, types of lighters, some hair dryers, medical supplies, fireworks, wheelchairs with wet cell batteries, breathing apparatus with compressed air, paint thinners, ammunition and pyrotechnics, among many others. Radioactive materials constitute a particular hazard, in that they emit particles (rays or gases), which may be hazardous and cannot be detected except by properly calibrated instruments. Radioactive materials offered for transport on passenger carrying aircraft must be intended for use in, or incident to, research or medical diagnosis or treatment.

Some hazardous materials or certain amounts thereof, are restricted to transportation in cargo aircraft only, and such materials should be stowed so as to be accessible to crew members while in flight. A logical but necessarily complex system was developed for labelling and describing dangerous goods, and great use is made of colours, shapes and symbols in order to contend with frequently encountered problems of language and illiteracy.

The ICAO system classifies Dangerous Goods into nine groups, with three packing group numbers:

- Packing Group 1 = great danger
- Packing Group 2 = medium danger
- Packing Group 3 = minor danger

In the event of an in-flight emergency, the aircraft commander is required to inform the appropriate agency of the class of dangerous goods on board the aircraft, so that the emergency services may be adequately prepared. He or she is expected to ensure that regulations for the safe carriage of the materials are observed, and he or she has authority to delay flights if necessary to ensure the proper stowage, quantity and documentation of shipments, and to refuse carriage of any dangerous goods that are not in compliance with the regulations.

'Airlines are required to ensure that all personnel, including pilots, involved in the acceptance, documentation, handling and transportation of dangerous goods, are initially and recurrently trained to appropriate standards. They must ensure that those items accepted for air transport meet the governmental regulations in force, and that the pilot is informed of any such materials on board his or her aircraft. Airlines are required to display in a prominent place, visible to passengers, notices concerning the requirements and penalties associated with the carriage of dangerous goods.'

The FAA is responsible for safety oversight of thousands of shipments of dangerous goods (hazardous materials) each day in USA civil air transport, and in 1993 made:

- 6413 operator/aircarrier facility inspections
- One freight forwarder inspection
- 22 240 individual shipments/document inspections
- Investigations of 489 discrepancies resulting in 563 enforcement investigative reports (there were 812 enforcement investigative reports in 1992)
- More than $1 million collected in fines

Shippers, packers and forwarders are required to properly package, identify and document all items. In a perfect world these precautions and the use of properly designed and equipped aircraft, would come close to guaranteeing safe transport of dangerous goods by air. Unfortunately in the real world there have been serious incidents and accidents.

Dangerous goods accidents

In 1987, a BAC 111 carried out an emergency descent when poisonous fumes affected the passengers. A passenger was prosecuted when investigation revealed that he carried ethyl chloroformate aboard in hand luggage, and that the containers leaked as the aircraft climbed.

In February 1988, an in-flight fire occurred aboard an MD-80, carrying 120 passengers and six crew near Nashville, Tennessee. The fire was caused by undeclared and improperly packed dangerous goods that leaked during transportation. The goods included five gallons of hydrogen peroxide, and 25 pounds of a corrosive material packed in a fiber drum and described as laundry equipment. Five minutes before landing, several passengers seated near the centre of the airplane saw smoke rising from the floor, smelled a burning odor and summoned a flight attendant. These passengers were moved to other seats, and two minutes after landing the captain ordered an emergency evacuation of the airplane. The airplane was badly damaged.

In February 1991, a DC-9-31 landed at Greensboro, North Carolina, taxied to the gate and when ground crew opened the cargo hold, a fire was discovered. The fire was extinguished, and the floor and wall of the hold had fire, heat and smoke damage, but the cause of the fire was not determined. There were 28 pieces of passenger luggage and six pieces of company material in the compartment. Inspections of the luggage revealed that two passengers had failed to declare dangerous goods, including a tear gas device, bottles of toxic dichloromethane, narcotic lamp oil, witch oil and safety matches. Had the fire reached these items of passenger luggage, the incident would have been more serious.

On a flight in 1993, a carrier found a partially destroyed mail sack that had contained bottles of hydrochloric acid used as toilet bowl cleaner. In yet another incident, mercury (a corrosive liquid), was found leaking from a mail bag during the loading of a jet airliner at Houston Texas. The shipper of the mercury had not identified its contents or warned of any hazard, and said that he had shipped mercury in this way for twenty years and was not aware that he had been doing so incorrectly. Many similar events from around the world could be listed in this catalog of near-disasters, and it is surprising that there are not more fatal accidents when dangerous goods regulations devised to protect the public, are breached.

In May 1996 a DC9 was destroyed in an accident over the Everglades, Florida. Preliminary reports indicate that the airplane was carrying 'company stores' that included airplane tires and up to 60 oxygen generators of a type used to supply passenger emergency oxygen masks. The oxygen generator canisters, normally containing sodium chlorate, were reported as being wrongly marked as empty. Retrieving wreckage from the swamp to establish the cause of the disaster was extremely difficult, but evidence was found indicating that there had been an explosion and resulting fire with thick black smoke. There was substantial evidence of a severe fire in the forward cargo area where the oxygen generators were stowed, and the cockpit voice recorder trace showed 'a brief unidentified sound' with the pilots reacting within 11 seconds and obtaining clearance to return to Miami within 28 seconds. The 'sound' is believed to be of an explosion in the cargo hold.

NTSB has recommended that air carriers revise procedures for accepting general cargo packages, to include questions developed to help identify dangerous goods shipped in packages and not declared. It pointed out that Department of Transportation (DOT) regulations require air carriers to inform passengers about dangerous goods by posting notices at locations where tickets are issued, but these are often posted at locations not readily visible to passengers. A third suggestion made was that the USA postal service improve its hazardous materials information program.

Particular safety problems involved in the use of 'combi' (passenger/cargo) airplanes were revealed in 1987, when a B747 'combi' crashed into the Indian Ocean during a night flight from Taiwan to South Africa, killing 150 people. Investigation showed that the design of the above-floor cargo holds was deficient, as in-flight fires could not be extinguished, and flight crew only had limited access for fire fighting, and the products of combustion were likely to affect passengers. In 1989, FAA and other regulatory authorities applied new and improved regulations to 'combi' airplanes. Among the required improvements are fire-lining for the above-floor cargo holds, improved fire detection systems, greater capacity fire extinguishers, improved training in fire fighting techniques for flight crews (some authorites require carriage of a designated fire fighter), frequent in-flight inspections of cargo, and improved access to cargo for fire fighting purposes.

In 1992, an MD-11 'combi' airplane was the first to be certificated with a main deck compartment, meeting the FAA's Class C fire and smoke containment requirements. This makes it possible to detect a fire within one minute, allow crew to activate the fire extinguishing system remotely (in the MD-11's case from the flight deck), shut down air circulation to the cargo compartment and to have extinguishant sufficient to fight a fire for three hours (an ETOPS requirement).

The presence of dangerous goods on aircraft complicate emergency procedures should there be an accident during take-off or landing. Rescue attempts and fire fighting are inhibited if the rescue services have first to deal with hazardous materials such as corrosive liquids and radioactive substances. Proposals to notify the carriage of hazardous materials to affected air traffic control units by adding an item to the transmitted flight plan of the aircraft, have not yet been accepted. The onus to notify the appropriate authorities in an emergency, has been passed to the pilot, who may have more urgent duties in these circumstances.

Millions of packages are airmailed each year by the general public, who are largely ignorant of the regulations that apply to the carriage of dangerous goods. Control of this traffic lies with the Universal Postal Union (UPU), representing the interests of state postal authorities. Its safety programs protect the safety of postal workers, and little is known of its methodology except that in times of high risk at a particular location, it may delay the carriage of mail for one or two days. It also uses pressure chambers in an attempt to defeat timer or barometric mechanisms mailed with explosive substances, and used by saboteurs to destroy aircraft in flight. Its education of the general public seems to be confined to the display of posters and other materials.

As so little effort is expended by the UPU in informing the general public about the risk to air travel posed by using air mail to send dangerous goods around the world, it seems that this particular risk will be present for many years to come.

Airlines display notices informing passengers about items that are not permitted to be carried as baggage either in the holds or passenger cabin, but difficulties continue to be experienced. This is perhaps because the notices are unseen until the passengers' bags have been packed and there is great resistance to leave behind any item. It is therefore not surprising that hazardous materials are found in the possession of passengers, when hand baggage is the subject of security screening at an airport.

The UK CAA has reported that in only one year, the following incidents arose as a result of dangerous goods items in passenger baggage:

- A chain saw leaking gasoline was found in a bag in an overhead locker
- A camping stove containing fuel was found in cabin baggage
- Baggage contained a hand-held smoke gun, five smoke canisters and leaking containers of butane
- Cargo hold baggage contained two liters of gasoline
- Baggage contained a canister of methylated spirits
- A child was showing fireworks to another child in flight
- Three bags were found to contain 40 kg of rifle ammunition

An unexpected feature of reported incidents is the large number of occasions when professional and technically qualified persons break the regulations. In one case, a rocket engineer travelling as a passenger carried a rocket motor containing explosives. In another incident, a passenger caused $12 000 of damage to the carpet in a terminal building while carrying a leaking jar of corrosive substance from one check-in counter to another. Other transgressors are hunters carrying ammunition in their luggage, which is often home-made and contains much more explosive than is normally commercially available. It is difficult to see an ending to this serious threat to air safety as airlines seem unable to find a solution to the problem.

As current trends are to relax ICAO Technical Instructions for the Safe Transport of Dangerous Goods by Air, and to allow more 'exempted' dangerous goods on board aircraft, risk seems likely to increase. These 'small quantities' of 'exempted' dangerous goods will *not* carry normal labels and the pilot-in-command will not be aware of the presence of dangerous goods that may be mixed in with other non-dangerous, as well as with dangerous goods. There will therefore be a reduced level of safety that will require the increased awareness of all persons concerned, if risk is to be contained at an acceptable level.

Equipment capable of detecting very small quantities of hazardous materials by means of 'super sniffing' techniques is becoming available, albeit at high cost. Cargo or baggage loaded in containers or igloos, is passed through the installed 'super sniffing' equipment at a rate of up to twenty containers per hour, and it is claimed that the equipment can detect minute quantities of any material for which it has been designed and programed. Air samples are gathered from the containers and passed through a test section containing a special disposable cartridge. Detection is based on the different molecular weights of the aromatic components of the target material, and it is claimed that even if explosives or drugs are tightly enclosed in film wrap, sufficient odours will be detected to raise the alarm. Until such equipment is in widespread use, it is likely that the dangers posed by dangerous goods will continue for a long time.

A further example of 'dangerous goods' incidents occur when passengers become disorderly, usually as a result of alcohol and/or drugs. There have been a number of occasions when pilots have made unscheduled landings to obtain assistance from police officers.

Accident prevention

Accident investigation is an important means of accident prevention, but by its nature is possible only *after* the event, and has to be augmented by investigating some incidents and obtaining safety-related information that would not normally be reported. Since the most important sources of safety information on flight operations and human factors matters are flight crews and air traffic controllers, it is necessary to devise schemes that provide anonymity to persons making reports. This is in order to ensure that the flow of information is not reduced because of fear of criticism from colleagues, disciplinary action by the employer or more serious action being taken by the state licensing authority.

Incident reporting systems

The first scheme of this type to be brought into operation was the USA's Aviation Safety Reporting System (ASRS), operated by the National Aeronautics and Space Administration (NASA). Started in 1977, the scheme attracted more than 35 000 reports during the first seven years of operation, and measures taken to ensure anonymity work well. The only circumstances where anonymity is not provided are when an illegal act is deliberately committed, a report is not made within a specified time scale or an accident has occurred. On all other occasions, the person making the report may be completely frank about his or her own and other persons mistakes and be critical about all aspects of the occurrence without fear of adverse consequences arising. The scheme identifies previously unknown problem areas and makes it possible to correct potentially hazardous conditions.

Many reports concern failures of the air traffic control system, particularly near-misses and failures to maintain assigned flight altitudes. This information would not be available had anonymity not been assured, for the USA's FAA is charged by Congress with enforcing Federal Aviation Regulations in addition to providing air traffic services. A report of an airmiss made prior to the introduction of the ASRS scheme, would have resulted in either the pilot or air traffic controller concerned being subject to enforcement action, for it was assumed that the system was perfect and any error therefore being the result of a lapse by the pilot or controller. Pilots and air traffic controllers were reluctant therefore to make reports and it was only after the scheme was introduced in 1977 that the FAA became aware of many failures of its own systems.

In 1993, the FAA asked the National Academy of Public Administration to review ASRS, and its report was published in August 1994. Although recommendations for improvement in administration were made, the report stated:

'ASRS has contributed to aviation safety, and human factors research and engineering for almost two decades. It has contributed to aviation safety by collecting, analysing, and communicating unique and valuable insights on incidents, which led to important changes in both processes and procedures. The Academy study team and advisory panel conclude that ASRS is a credible, resilient and worthwhile program that adds to our understanding of the human elements of the aviation transportation arena.'

In 1993, 30 498 reports were filed. Air carrier pilots filed 20 659 reports (68%), general aviation pilots 8093 (27%), controllers 1356 (4%), and others (flight attendants and ground crews), 156 (0.5%). ASRS data is used as background information for human factors studies in other industries where safety is particularly important, and numerous requests for information are received from foreign nationals and airlines. ASRS is therefore a great tool for USA and international air safety!

In the United Kingdom, a similar scheme was started in 1982 and is referred to by the acronym CHIRP for Confidential Human factors Incident Reporting Program. Confidentiality was ensured by placing administration of the scheme in the capable hands of the Royal Air Force's School of Aviation Medicine with the principal terms of reference being to:

■ Receive incident reports and ensure, by any necessary immediate follow up, that as complete a picture as possible is obtained
■ Initiate any necessary urgent action by alerting the appropriate body after first effectively disidentifying the incident report
■ Effectively disidentify all other reports, put them into a standard format and pass them on to the safety data unit
■ Ensure rapid feedback to individual reporters

In 1994, CHIRP was reviewed and changes were made in its administration with representation on its management board for medical experts, pilot organisations, airline operations directors and aerospace manufacturers. Financial support is from the UK Civil Aviation Authority and assurances were given that its objectives remain unchanged.

Countries in Europe are to form a similar organisation: European Confidential Safety Reporting network (EUCARE), that hopes to become a coordinated system for 15 countries in the European Union. Eight countries participated in a steering group that has agreed that the Technical University of Berlin should operate a central data base when the network becomes operational in 1997.

In the USA, an American Airlines initiated aircrew 'no blame' incident reporting system increased the number of reports by a factor of more than 50, and has the blessing of the FAA and the Air Transport

Association. In mid 1995, it seemed likely that other airlines would join the program which is seen as a supplement to ASRS, with reports sent to ASRS after being 'de-identified'.

Other safety programs

The International Civil Aviation Organisation, ICAO, has an accident/ incident data reporting (ADREP) data bank, that is a prime source of information on reportable accidents/incidents that occur in the 180 plus states who are members of the organisation. All reports made since 1970 are stored in a computer, and are available to any member state or other organisation recognised by ICAO. The total number of accident and incident reports stored by the end of 1995 was 18 759 with 1405 reports received that year. The way in which data is stored makes it possible, for example, to retrieve all reports of landing accidents to a particular aircraft type where icing conditions are believed to be a factor. Trends are identified and this data has proved to be of great value.

In 1976, the UK Mandatory Occurrence Reporting (MOR) scheme was introduced. The scheme has two principles. Firstly, it includes in one system the occurrences relating to any aspect having a direct bearing on flight safety, e.g. flight operations, aircraft airworthiness, air traffic control and associated facilities. Secondly, it monitors those occurrences which result in, or could result in, a significant hazard. The objective is to use reported information to improve the level of safety and not to attribute blame. The word 'occurrence' includes accidents and incidents, and the reporting of occurrences is mandatory for all UK public transport operations. Reports of information gained are free to participating organisations and are widely circulated on a subscription basis to other parties. The computer and data base is compatible with ICAO's ADREP and those of other regulatory authorities, and more than 5000 reports are received annually.

The USA's FAA system of Service Difficulty Reports (SDRs), has objectives similar to that of the UK's CAA MOR, but it is more limited in its application, and the FAA is unable to keep information confidential due to Federal Laws that provide public access to information.

In early 1995, agreement between the FAA, the USA airline industry and airline pilot organisations, called for the development of Flight Operations Quality Assurance (FOQA) programs. The programs, which are in widespread use in Europe, are designed to use digital flight data recorder (DFDR) parameters to monitor operations for unsafe or unauthorised practices, and to allow airlines to implement corrective actions to prevent incidents and accidents.

By November 1995, United and USAIR had started their own programs. At least six other airlines were almost ready after they and their pilots had reached agreement over data security, and the airlines and the FAA had negotiated an agreement that all information would remain on airline premises and would not be given directly to the government. A typical use for FOQA would be to detect actual, or near, tailscrape incidents and to examine all relevant parameters so as to devise better piloting techniques and thereby improve training. The program is expected to detect safety trends before they become incidents or accidents, and to devise and implement improved procedures. British Airways was able to implement the first scheme 30 years ago because the UK regulatory and legal systems permitted the confidentiality necessary to gain their pilots' acceptance. The 'spy in the sky' becomes the pilot's friend with the introduction of these programs!

The FSF acting as a consultant organisation working for fees, carries out safety audits for client airlines. It states that:

'an aviation safety audit is a procedure whereby designated and qualified persons systematically and objectively look at the activities of an aviation organisation in the context of that organisation's own operating plans as well as in relation to industry practices and applicable government regulations . . . it is more commonly recognised as providing valuable information for management to act upon in the interests of improved safety.'

An important audit activity is to conduct personal and confidential interviews with employees at all levels within the organisation. The following are example comments arising from safety audit reports:

- The CEO was unaware that he was personally contributing to a lower margin of operational safety by his favoritism and tacit acceptance of non-standard operating procedures
- The President (of a small airline) consistently injected himself into situations involving line maintenance personnel, and in some instances even changed the instructions or work projects previously issued by the assigned supervisor
- Department heads or middle managers have been observed circumventing the first level supervisor. Bypassing the chain of command causes confusion and frequently disrupts planned activities, or work in progress.

In mid-1996 the FAA proposed that a 'workable global safety database' be developed and operated by the private sector to contain massive amounts of aviation safety information. The proposed system would collect and analyze safety data to determine trends and frequent occurrences, with this derived information being available without restriction. Information would be volunteered by all sections of the aviation community and private ownership and operation is favored because it would remain clear of the Freedom of Information Act. It is possible that the FSF would operate the proposed system.

Aircraft manufacturers' safety programs

The Boeing Commercial Airplane Company has a great record of participation in safety programs as shown by its work in Society of Automotive Engineers (SAE) committees, and in the ICAO/FSF program to eliminate CFIT accidents. Its surveys of accidents and trends are widely used by others, and the magazine 'Boeing Airliner' is a valuable source for all students of air safety, with the company participating in many meetings that set out to discuss and analyze safety problems. It would be difficult to write a book on air safety without using Boeing data.

Airbus Industrie has a voluntary confidential reporting system for operators of its airplanes, because airlines were not using the International Air Transport Association's (IATA) safety information exchange for fear of being exposed to litigation. Manufacturers are usually told quickly about technical problems with their airplanes, but are not always informed about operational incidents and this is particularly true when events are purely operational. Operational incidents are important because they may reveal a failure to follow standard operating procedures (SOPs) or crew difficulties experienced in dealing with a problem, and this may lead to an improved product. By early 1995, 34 airlines had joined the program, representing 23% of Airbus operators but covering 50% of airplanes in operation. Reports are de-identified and entered into a data base and all information sent to participating airlines is to be used only for the sole purpose of accident prevention.

Aircraft systems and equipment for the prevention of accidents

There are occasions during a take-off when a pilot may doubt that the airplane can safely continue because of a worse than expected airplane performance for reasons of a dragging brake, snow or slush on the runway, an overloaded airplane or engine malfunction etc. Should the pilot decide to abandon take-off, there will almost certainly be an incident with tire bursts and wheel fires and possibly a runway overrun and serious accident. Research into take-off performance monitor systems holds out the promise of taking guesswork out of pilot decisions at such times. As NTSB reports show, more than 4000 take-off related accidents between 1983 and 1990 resulted in 1378 fatalities, and the need for improvement in safety during this phase of flight is obvious.

A NASA report in 1994 holds out the promise of it being possible to measure along-track acceleration relative to a computed nominal acceleration based on existing conditions and a standard (ideal) execution of the take-off maneuver. A NASA program investigated the potential of a Take-off Performance Monitoring System (TOPMS), that uses software to drive cockpit displays that graphically indicate take-off performance relative to a reference performance, engine condition, and a continually updated prediction of the runway position where the airplane can be braked to stop if an aborted take-off becomes necessary. It also provides explicit 'go/no go' advice in the form of situation advisory flags. Flight tests using 32 research pilots (US Air Force pilots and professional civilian pilots) have shown that the displays are easy to monitor and provide valuable safety, performance and advisory information currently unavailable in commercial cockpits. NASA concluded that:

> 'it is desirable to demonstrate and evaluate TOPMS on another airplane with different characteristics, sensors and support equipment (e.g. a head-up display). The TOPMS software could easily be adapted and used to advantage on any modern airplane equipped with a digital flight-control system.'

Because airplane accidents have been caused by total flight-control systems failures, including a DC10 at Sioux City, Iowa in 1989, NASA has been investigating propulsion-controlled landings using an MD-11 test-bed with modified flight control software. The MD-11 used, has standard hardware with software modifications to number one flight control computer and the engine controls. The standard auto flight system heading/track knob, and vertical speed/flightpath angle thumb-wheel are used for the Propulsion Controlled Aircraft (PCA) program, and successful 'manual' and autopilot-coupled ILS approaches have been flown to touchdown with all hydraulic systems switched off. Other types of airplane have been examined, and it has been concluded that they all have adequate propulsion control capabilities for extended cruise flight *and* that every airplane type is very difficult to land using manual throttle control. Software-commanded differential power from the three engines provides changes in pitch, bank and yaw, and it was concluded that the potential for backup propulsion-based flight controls on new designs was promising.

9: Accident Investigation

ICAO Standards and Recommended Practices

Article 26 of the Chicago Convention of 1944, applicable to all 180 plus member states of the International Civil Aviation Organisation, imposes an obligation on the state where an accident occurs to institute an inquiry in defined circumstances, and as far as its laws permit, to conduct the inquiry in accordance with ICAO procedures. It also entitles observers from the state in which the aircraft is registered to be present at the inquiry, even if that state is not the state in which the accident occurs. Article 37 of the Convention specifies that the Standards and Recommended Practices for aircraft accident inquiries are to be produced in Annexe 13 to the Convention. The articles and the annexe are designed to provide a common international framework for the investigation of accidents in civil aviation. It was foreseen that an aircraft involved in an accident may be registered in country A, constructed in country B, operated by an airline in country C, flying between countries D and E and suffer an accident in country F. Annexe 13 states that the objective of an accident inquiry shall be the prevention of accidents and not to allocate blame or liability.

Although some states conduct inquiries with that objective in mind, widely differing national laws and levels of technical expertise and resources can result in the publication of accident reports falling short of the desired standards of objectivity. What are perceived to be national and public interests in aircraft accidents tend to interfere with the pursuit of the ideals expressed in the Chicago Convention and Annexe 13. In many countries the airline and major airports are owned and operated by the state, and the air traffic control agency is similarly a national agency. A conflict of interest can therefore arise when an accident occurs and a state carries out an investigation and inquiry which may reveal that its own agencies are culpable.

In 1992, ICAO proposed 33 amendments to Annexe 13 following its first meeting to address accident investigation problems since 1979. The meeting considered flight recorders, accident/incident reporting systems, training seminars on accident investigation, investigation of accidents to leased, chartered and interchange airplanes and the role of human factors in accident investigation.

National practices

Some states ensure impartial investigations and inquiries, by forming independent accident investigation bodies. In the USA, it is the law that National Transportation Safety Board (NTSB) accident reports cannot be used in damage suits arising out of the accident, and its reports are occasionally critical of the Federal Aviation Administration. In the United Kingdom, the Air Accident Investigation Branch of the Department of Trade is independent in that it reports directly to the Minister of the Department, can criticise the Civil Aviation Authority and can recommend changes in regulations and procedures.

These high standards are not achieved in all countries and particularly in the so-called developing world. As a result, there have been cases where a state considers an accident report published by another state to be so seriously deficient that it publishes its own report or a formal comment on the original report. In 1979 the ICAO Annexe 13 was amended so as to permit an accredited representative of another state to append a dissenting minority attachment to the official accident report. The United Kingdom used that right in 1980 to append an attachment to the Spanish report of an accident to a British registered Boeing 727 that crashed at Tenerife in April 1980.

Accident investigations

States with major involvement in the civil air transport industry have teams of trained investigators on standby at all times. Their investigation kits, including technical measuring equipment, cameras, protective clothing and other clothing suitable for wear in any climate, are always ready for use.

In the United Kingdom, a decision on whether an investigation is to result in a Public Inquiry is made by the Secretary of State, and the determinant is usually the degree of public concern and national interest. Increasing technical complexity of aircraft operations and accident investigation is tending to favor the non-public forum, but in every case the findings are promptly sent to all interested technical parties so that any preliminary and precautionary actions needed may be quickly taken.

Since 1969, any person whose reputation is likely to be called into question is provided with a copy of the final draft report and is allowed to make representations or to seek changes before publication. This procedure has resulted in changes being made to final reports where, for example, a pilot made an error but the initial report failed to take account of extenuating circumstances. As a result of an abuse of this procedure by an airplane manufacturer who delayed publication of a report for two years, a limitation is now placed upon its use for 'frivolous purposes'.

Procedures in the USA are very different, where the Freedom of Information Act leads to expectations of an early public hearing where a number of persons and organisations are party to the proceedings. This is where ground seeds for later litigation are set, perhaps to the detriment of the truth, because witnesses and investigators are put under great pressure to provide sought after answers.

A cause for concern has been the tendency of the NTSB to publish a report that includes a statement, 'NTSB determines that the probable cause of this accident was...'; because its mandate from Congress is to 'determine the probable cause(s) of accidents'. A growing number of accident investigators and aviation safety experts believe that ascribing an accident to a single probable cause is unproductive, and leads to focus on the 'who' and 'what' of an accident rather than on the 'why'. Studies of real life accidents usually show that a chain of events leads to an accident and that to focus on the final link – that is probably the failure of flight crew to interrupt the chain – prevents safety lessons being learned. An examination of tables of accidents in Chapter 8 shows that almost every accident has more than one causal factor.

In the final report of the 1992 ICAO accident investigation meeting, it is stated that causal statements should:

- List all the causes, usually in chronological order
- Be formulated with corrective and preventive measures in mind
- Be linked to related and appropriated safety recommendations
- Not apportion blame or liability

and it is to be hoped that all member states of ICAO will adopt these recommendations, but in the USA a change would need to be accepted by Congress.

The objectives of on-site investigations are similar everywhere. They are to avoid early interpretations taking the form of a theory that could prejudice the importance of any information that becomes available at a later stage, and to seek corroboration for all items of evidence. Ideally, all pieces of wreckage are recovered after their position in the pattern of wreckage has been plotted, and all witnesses, survivors and flight crew are questioned. Post mortems are held and victims are identified against their known seating positions. The recovery of flight recorders takes a high priority and investigators take every possible step to prevent them being impounded by coroners, police, and investigating magistrates. It is seen as a vitally important task of accident investigation to ensure that causes are known as soon as possible, in order that immediate action may be taken to prevent any other aircraft suffering a similar accident. Delays in obtaining flight recorder information could prevent that objective being achieved.

Failures of structures, materials, engines, flight control systems, aircraft systems, human failures and combinations of all these failures are looked for. Air traffic control radar plots, tape recordings of radio transmissions and reports of the weather prevailing at the time of the accident are also examined if available. Some of the work of investigators is similar to detective work, particularly when sabotage or a failed attempt at a hijacking is suspected, and the help of forensic and explosives experts is sought if needed. The work is difficult, tedious, slow, uncomfortable and unpleasant, being sometimes carried out in remote places in severe climatic conditions.

When an on-site investigation is complete, wreckage may be moved to a place where better technical facilities are available and where all the pieces can be set out in the correct position relative to each other with further exhaustive tests taking place until a complete understanding and appraisal of the evidence is available. These efforts cost a great deal of money but are accepted as essential if air safety is to be maintained.

A possible way of avoiding problems of technical incompetence or shortage of financial resources would be to make accident investigation an international activity under the auspices of ICAO, or alternatively of regional organisations such as of the European Community. Assurance of objectivity, sharing of costs and pooling of expertise would be three advantages of such arrangements, and although adoption of the idea is not imminent, the possibility offers a chance to achieve improved air safety.

The trend towards globalisation in the aviation industry poses a new challenge for accident investigation authorities. Components or parts for an aircraft, including engines, are manufactured in three or four countries and assembled in yet another. The aircraft may be registered in one country, leased to an airline in another which in turn has maintenance, overhaul and baggage handling arrangements with companies in several other countries. Owing to operators' alliances with other airlines, it may be difficult to determine where operational control and responsibility for safety lies.

It has been suggested that there is a need to establish an international independent aircraft accident investigation agency, perhaps because two recent accidents needed multi-national investigation teams. One airplane was of Thai and the other of Pakistan registry and both accidents happened at Katmandhu, Nepal in 1992. If such an international authority were to be formed it would almost certainly be under the auspices of ICAO, and until that day comes there is a need for improved inter-communication and interaction between national authorities.

A recent development in accident investigations is to examine what are sometimes called 'Management Factors'. This involves an examination of the part played by management's policies on the areas of safety regulation, airline operations and manufacturing. It is easy to identify personnel directly involved in an accident – pilots, maintenance engineers and mechanics and air traffic controllers, but more difficult to establish a causal factor arisng from acts of commission or omission by managers. An article in the October 1995 issue of the ICAO Journal reviews an official report of an accident in Canada, and sets out to consider the influences of all the organisations involved in the operation as well as individual human performance. It identifies what are termed 'error producing conditions' and 'latent organisational failures' and produces a 'causal pathway' (shown here vertically) that starts with:

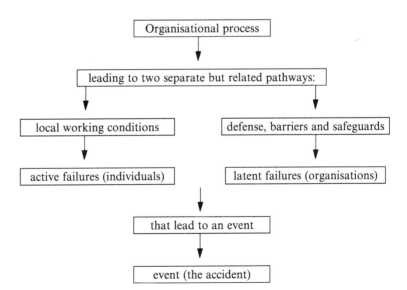

The authors conclude that 'arguably, many unsafe acts were none other than the behaviours fostered by the aviation system itself'. The article is a good illustration of an important trend in the search for air safety.

Accident investigators

A welcome feature has been the introduction of accident investigation courses at a number of technically competent seats of learning. Courses are available at the University of Southern California, Cranfield Institute of Technology in England and the Institute of Aviation Safety of Stockholm and a high standard of training for accident investigation is achieved.

In 1979, the USA NTSB closed its accident investigation school at Dulles Airport, Washington because of financial constraints, but the decision was reversed and a new school opened at Oklahoma City in 1985. The reasons given for the changed policy include an increased use of new materials in aircraft construction, electronically controlled flight control systems that tend to make older investigative techniques obsolete and the use of electronic flight instrument displays 'that disappear like magic in an accident'. The course provides students with an ample opportunity to gain 'hands on' experience in typical wreckage layouts, and studies include witness interviews, mid-air collision case studies, material failure analysis, in-flight break-up cases and postmortem examinations etc.

The International Society of Air Safety Investigators (ISASI), holds annual conferences to study improved methods and technology. Participants include airline pilots, representatives of aircraft and equipment manufacturers, airlines and state authorities such as the USA's NTSB. Pilots point out that newly available investigative techniques can invalidate findings of earlier accident inquiries, and argue that some earlier accidents assigned to pilot error were almost certainly caused by windshear. A similar argument is applied to accidents of some 30 years ago before the phenomenom of aquaplaning was understood, when aircraft inexplicably overran runways on landing with the pilot usually being blamed for failing to carry out standard operating procedures.

Flight recorders

Early flight recorders performed only an 'eyewitness' function, recording five or six parameters by making scratches on a metal foil medium.

Parameters were recorded in analog form and were usually altitude, airspeed, magnetic heading, time, and *g* forces. The recorded information was useful where no reliable human observer had witnessed an accident, or to supplement reports made by humans, but it was obvious that additional valuable information could be provided by the use of improved flight recorders. It was appreciated that the recording of additional parameters would enable valuable lessons to be learned from the investigation of *incidents* in addition to accidents, with the generally accepted definition of an incident being 'an accident that almost happened'. Other uses foreseen for the increased amounts of information that would become available, was the monitoring of the performance of the aircraft and its engines and, perhaps, also the monitoring of the performance of flight crews.

These concepts have been developed in the form of three separate recording devices named: Flight Data Recorders (FDRs), Cockpit Voice Recorders (CVRs) and Maintenance Recorders (MRs). To the general public and journalists they are known as 'black box flight recorders' although they are usually bright orange in colour to facilitate location at crash sites.

Flight Data Recorders (FDRs)

A flight data recorder was used by the Wright Brothers on their first flight, recording engine revolutions, distance flown through the air and duration of flight. It was many years before airlines followed that excellent example, and many more before internationally agreed standards were set by ICAO.

In 1983, ICAO reviewed its ten year old standards for flight data recorders. It decided to update the minimum requirement of having five-parameter recorders for older turbine-engined airplanes, with a more demanding requirement for wide-bodied aircraft. This was because of the vital role of accident and incident investigation in promoting safe and efficient aviation. The ICAO Bulletin for October 1983 stated:

'With the escalating costs of accidents, it has become increasingly important to ensure that anything which might have a bearing on the safety of aircraft operations be determined at the earliest possible stage of an investigation. To achieve this, certain improvements in equipment and procedures are needed to gather and analyze accident and incident information rapidly ... The continuing evolution of modern aircraft has placed added emphasis on the capabilities and limitation of the human being in the overall man–machine environment system. The techniques of aircraft accident investigation must also evolve if investigation is to remain an efficient means of accident prevention.'

The article went on to state that digital avionics have outmoded analog mechanical systems, expanded-parameter flight recorders are the only viable means of acquiring vital information, and metal-foil type FDRs do not provide sufficient data to adequately determine what occurred in an accident. ICAO therefore recommended the introduction on or after 1 January 1987, of Type 1 digital FDRs capable of recording 33 parameters for at least the last 25 hours of aircraft operation, or Type 2 digital FDRs capable of recording at least 15 parameters for a minimum of 30 minutes, with the choice of type of FDR determined by the type of aircraft to which they would be fitted.

In mid-1987, the USA required that all large transport aircraft have their old-type foil recorders replaced by digital recorders (albeit with only six parameters being recorded), with only newly manufactured aircraft being required to have 17 parameters on the recorders. Safety experts welcomed these improvements but deplored the limited number of parameters to be recorded, particularly on the older aircraft types. The NTSB, which led the campaign for better FDRs, stated that a review of its accident/incident files for 1983/1986 showed a high failure rate for metal foil type recorders, and that at least 48% had one or more malfunctioning parameters preceding the incident/accident, preventing the read-out of pertinent data. It described the new requirements as being 'too little – too late'.

Cockpit Voice Recorders (CVRs)

With regard to Type 1 CVRs, ICAO concluded that the continuous loop recording time should be increased from 30 to 60 minutes and the number of tracks recorded should be at least four. For simple or smaller aircraft types it was recommended that Type 2 CVRs should record at least 2 tracks for at least 30 minutes.

Maintenance Recorders (MRs)

There is no mandatory requirement for the installation and use of Maintenance Recorders, but some airlines have installed them to record as many as 200 parameters, mainly for the purposes of monitoring the performance of aircraft, engines and systems. In some cases MRs have been combined with FDRs, with only the parameters required for the purposes of accident and incident investigation being 'protected' in the accident case. A new trend is for an airline's maintenance base(s) to be able to receive data from aircraft away from base, either by transmitting data over telephone or telex lines from intermediate points of landing, or even by real-time radio links from the aircraft while in flight. With a two-pilot crew becoming standard, MRs replace the data recording function of flight engineer/third pilots.

A combined FDR/CVR/MR is technically possible but operationally undesirable. Concern has been voiced by accident investigators at the practice of extracting data from FDR recordings for maintenance purposes, because a possibility exists of corrupting recordings that may be vital to the purposes of the accident investigators in a future accident that may occur on a subsequent flight.

Flight recorder location devices

To facilitate the location and recovery of flight recorders, ICAO recommends that they be painted a distinctive colour and carry reflective material. It also proposed that they be fitted with an automatically activated underwater locating device which, it was noted, had been proven to be of great practical value in some accidents, and were already required equipment in some states. Available types of location devices include water, or *g* activated radio beacons, noise-makers, or atomic trace elements. The location devices require their own independent power sources that should last for at least 14 days.

Examples of the operational requirement for improved underwater location devices are: the search for the flight recorders of the Korean Air Lines Boeing 747 shot down by fighter aircraft of the Soviet Air Force in September 1983, and the retrieval of flight recorders of the Air India Boeing 747 that crashed in 6700 feet of water off the coast of Ireland on 23 June 1985.

Airline use of FDRs and CVRs

It is arranged for the FDR and the CVR to be functioning at all times that an aircraft is in motion under its own power. Flight deck crew are not permitted to switch off the equipments, except at the end of an incident-free flight. The CVR operates continuously on a 30 or 60 minute 'loop' with every item of audio information including radio contacts, cockpit conversation, and audio inputs and outputs to and from all of the radios being recorded. Typical airline use for an FDR is to record:

- Time
- Altitude
- Speed
- Course
- Flight attitude
- *g* forces
- Engine power
- Operating conditions such as rpm, temperatures and pressure ratios
- Flight control positions and pilot inputs
- Aircraft configuration such as wing flap and landing gear positions
- Automatic flight control system usage
- Radios in use and the performance of all safety critical systems

Pilot associations have cooperated well with airlines and authorities in what could be regarded as 'an invasion of privacy', with the basis for most agreements made between airlines and pilot associations being that the information stored in the FDRs and CVRs is used only in the interests of air safety. In a number of countries (after agreement between airline, regulatory authority and pilots), FDRs are used to detect 'events' in flight operations, usually by indicating when an envelope surrounding an individual parameter is exceeded. As an example, it may be agreed that the maximum normal pitch angle on take-off is between 12 and 14 degrees, and that the envelope should be set for 16 degrees so that rotation beyond this figure would trigger an 'event'. FDR recordings are routinely examined by automated programs and arrangements are made for the pilot to provide comments or explanations (on a no-fault attribution basis) to gain a better understanding of why an event occurs. In some cases pilot training programs are amended after events show a requirement so that all airplanes are operated in a standard manner likely to reduce the possibility of incidents/accidents such as 'tailscrapes' on take-off. These programs were introduced in the USA in 1996 and are called Flight Operations Quality Assurance (FOQA) programs, see chapter 8.

FDRs/CVRs in accident investigation

In the event of an accident or incident, information stored in the recorders is vital evidence. It is normally in the care of the chief accident investigator so as to provide security and privacy for vitally important flight safety information. These arrangements work well in most cases, but are threatened in countries where an investigating magistrate or coroner may gain early possession of the information, and because of a lack of appreciation of air safety matters, misinterpret or even publish the information. Journalists quote CVR information out of context and 'try' the flight crew in public even before the official investigation and inquiry has begun.

For a number of years the Head of the United Kingdom's Accident Investigation Branch (AIB) was Mr Bill Tench, and in his book *Safety Is No Accident*, he writes:

> 'The truth of the matter is that the FDR provides the investigator with a very useful tool which can produce a reliable and totally impartial record of the primary flight parameters against a time base, from which it is possible to reconstruct the flight path in three dimensions. The CVR allows the investigator to know what the members of the flight crew were saying to each other, and to some extent what they were thinking, at relevant times during the flight. It also serves the very important purpose of providing the key for synchronising the FDR and CVR time bases with ground based recordings of the radio communications.'

Mr Tench's views are supported by pilots, particularly on the value of the impartiality of the data recorded. They believe that earlier use of flight recorders could have prevented evidence given by pilots being discounted, because it did not accord with an existing state of technical knowledge and theory. They also believe that problems that caused accidents and took lives before the 1960s when flight recorders began to come into use in the airlines, would have been recognised and overcome if flight recorders had been available. Examples of safety advances that could have been made earlier if comprehensive FDRs had been available are: recognition of slush/snow drag effect on take-off performance, aquaplaning, and early jet pressure-hull problems.

There have been cases where data contained in flight recorders has been misinterpreted. It is generally recognised that the USA and the United Kingdom have the best technical resources and experience to interpret data that may have been damaged by exposure to the very high *g* forces and temperatures experienced in some accidents. Other difficulties are the occasional mismatch between the information displayed to the pilot in flight and that in the flight recorder, and a human tendency for investigators to note only that information that accords to their own most probable theory and to ignore data that does not fit.

Successes/failures in accident investigation

Design features and materials used in many types of airplane have been improved as a result of findings of accident investigations:

- Structural strength and take-off technique for early Comet airliners
- Engine pylons and fuse pins of Boeing 707, 757 and 747 airplanes
- Hydraulic systems and flight control runs of DC10s
- Anti-icing/de-icing systems of DC9s and several turboprop airplane types
- Autopilot design features
- Rudder trim controls of B737s
- Improved training in automatic flight control systems for pilots of Airbus A320s
- Propellors for Embraer Brasilias
- Reverse thrust systems (to prevent unwanted operation in flight)
- Flight instrument displays including altimeters
- GPWS and TCAS
- Improved repair schemes for ageing airplanes
- Modifications to the reverse thrust sensing systems of B737s after a number of landing accidents and incidents

Airport lighting systems and markings have been improved after accidents demonstrated a need. A preliminary report into a fatal accident to a Turkish registered B757 off Dominica in 1996 is likely to result in improved arrangements (better management) to protect aircraft systems during long periods on the ground, and better pilot training to recognise malfunctioning systems. (It is believed that the pitot heads were left uncovered for a period of several days, became obstructed and caused the captain's airspeed indicator to grossly over read. The pilots believed the displayed false reading and stalled the airplane in attempts to reduce the airspeed.)

One outstanding five-year-long failure (a mystery), is the cause of several accidents to B737 airplanes where rudder 'hard-overs' are believed to have caused pilots to lose control. The combined efforts of the manufacturer, NTSB, regulatory authorities and airlines have failed to find a cause. The NTSB stated that if all the airplanes had enhanced FDRs instead of six-parameter FDRs, the problem might have been identified earlier and perhaps solved. By mid-1996 the FAA had rejected a NTSB proposal that B737s should have enhanced FDRs fitted in a special high priority program and had instead issued proposals that all passenger airplanes be fitted with enhanced FDRs during 'heavy maintenance' over the next four years.

Technical developments

A new and probably extremely useful development in the use of FDRs studied in research establishments is the display of data in the format of normal flight instrumentation on monochrome cathode ray tube (CRT) displays. High speed computing capacity enables the display to give a picture of high realism, enabling the investigator to build up an appreciation of events in a manner similar to that of the pilot involved in the accident.

The use of electronic instrument displays in airline aircraft is a source of concern to accident investigators, for information on display is likely to disappear at the instant of impact whereas mechanical instrument pointers often leave tell-tale marks on the face of the instrument. A further possible source of difficulty is the multi-function usage of the new displays where the pilot chooses what is to be displayed. A choice of aircraft attitude, navigational or weather information, check lists, emergency procedures or synoptic displays of the aircraft and engine systems is available. An investigator may not therefore be able to establish what information was displayed to the pilot at the time of the accident. UK investigators have proposed that data supplied to the CRTs be recorded for the purposes of accident investigation, in order that they may know what was on each display at the time of the accident.

In the USA and UK, accident investigation authorities are interested in providing miniature charge-coupled digital cameras fitted under the aircraft nose to look back at engines and landing gear, with a dedicated display screen for the use of flight crews. It is believed that these systems would enable pilots to detect and identify serious problems where the sight lines and geometry of modern airplanes leave pilots 'blind'.

A USA specialist in airborne recording systems has proposed to Boeing and McDonnell Douglas the use of a Crash-Protected Video Recorder (CPVR), for electronic flight instrument displays and proposed external observation systems. The system would involve the integration of a CPVR through the central display electronics unit with the electronic flight instrument system, engine multi-function displays, external observation systems and cockpit audio sources. Recording times of up to four hours are envisaged and the system would withstand temperatures of up to 1000°C for up to ten minutes.

In Canada it is proposed to use FDR data to reconstruct the flight path of the airplane, synchronize it with the CVR recording and show the flight path with a data base of terrain, aircraft models, instruments, and instrument panels to create a dynamic real-time enactment of what may have happened.

At a 1982 meeting of the International Society of Accident Investigators (ISASI), the following question was posed:

'Why don't we, the aircraft accident investigators, access the brain centre of the airplane and record the action in the cockpit with a video recorder?'

It was suggested that the equipment be called a Control Cab Video Recorder (CCVR) and that it would be a perfect complement to the voice recorder. When combined with the Flight Data Recorder (FDR), a new dimension in aircraft accident investigation would be available and a CCVR would be a great tool to more positively determine the causes of accidents. Its use would remove much guesswork and doubt, and ultimately prevent future reoccurrences. It was stated that there is more information in the cockpit of an airplane than in any other place on board, and recording that information with a CCVR, for use only in case of an accident or incident, could be the single most important investigative tool yet conceived. It was pointed out that all voice communications in the cockpit are recorded, many parameters on the controls, engines, and flight attitude are recorded, but making a visual record of what was going on in the cockpit had been resisted.

Arguments put forward to support CCVRs included instances where the proposed equipment would have provided information on whether intruders were in the cockpit at the time of an accident, as has been suspected on at least two occasions. Their use would make information available on what was being displayed to the pilot at the time of an accident, crew inputs to flight controls, their management of the aircraft systems, and this information would be much more easy to retrieve than from FDRs.

During trials conducted by the Boeing Company some fifteen years earlier using a TV camera in a flight simulator, the images were so good that the wrist watch worn by one of the pilots could easily be read. Comment was made that the average person can retain 12% of an audio input for approximately 3 days; about 25% of a video input for the same period, and when both are combined, up to 65% is retained, and that all these factors taken together made a powerful case for CCVRs. Pilots would possibly accept visual monitoring of their work place if they were assured that systems would be used only for accident/incident investigation and that traditional methods of accident investigations would also continue to be used.

Accident investigation and criminal liability of pilots

Airline pilot perception of vulnerability to prosecution in criminal courts after involvement in serious accidents, is a factor that may affect their future cooperation in accident investigations and inquiries, and the continued availability of safety information important for the prevention of accidents.

A basic problem is legal liability of the pilot in the event of an aircraft accident. Aircraft accident investigation is provided for in Annexe 13 of the Chicago Convention which clearly states that the objective is the prevention of incidents and accidents and not to apportion blame or responsibility. Some member states of ICAO follow that concept but in other countries the investigation and following inquiry are regarded as a starting point for the apportioning of blame and subsequent litigation. Procedures followed can lend themselves to that practice, with technical evidence being challenged by persons who are not competent to make assessments as to its value, and criminal code legal procedures with their concept of prosecution, defence, hostile witnesses and guilt, taking precedence over finding the cause(s) of the accident and how to prevent a similar accident.

When crew members survive an accident their testimony is crucial to the task of finding the cause(s), their voluntary cooperation is relied upon for this purpose and they are interviewed during all phases of post-accident investigations. They are expected to cooperate even when they may have performed their duties imperfectly and started a chain of events resulting in an accident. In incident reporting schemes, they are expected to provide details of events that may be known *only* to themselves, and again where they may have failed to perform their tasks perfectly.

The only protection provided to information given by flight crews stems from paragraph 5.12 of ICAO Annexe 13. This provides a limited degree of protection in response to Freedom of Information legislation adopted by a number of states, that believe the public interest requires that accident information be freely published. It is this conflict of interest, between the right of the public to know about safety matters and the need of accident investigators and safety organisations to continue to receive cooperation and all available information from flight crews, that has to be addressed.

IFALPA representing more than 80 national airline pilot associations, is concerned that failure to protect the confidentiality of safety information will lead to flight crews refusing to provide information arising from incidents or accidents in which they have been involved, where that information may be used in legal proceedings against them. Pilots have reason for concern as shown by these occurrences:

- In Taiwan in 1968, two USA pilots were charged with manslaughter after a fatal accident and were facing possible sentences of two and five years imprisonment until international pilot pressures brought a retrial and acquittal on grounds of previously ignored technical evidence
- In 1979, two Swiss pilots were charged with manslaughter after a fatal accident at Athens airport. They were found guilty and sentenced to five years and two months imprisonment, and released only after an appeal to the Supreme Court based on technical evidence. This showed that the major causal factors of the accident were the many physical and operational deficiencies of Athens Airport. The trial had not taken this evidence into account
- In Venezuela in 1983 after a landing accident, two pilots were sentenced to 15 years imprisonment reduced to eight years on appeal, and they were subsequently granted a pardon by the President of Venezuela. This was after IFALPA assistance had shown that the airplane landing gear collapsed as a result of a fatigue crack
- In 1989, a local prosecutor in Queens New York, filed criminal charges against pilots involved in an accident
- In 1991, a British court considered charges brought by the CAA against a pilot who narrowly missed collision with an airport hotel during an approach to land in poor visibility. The pilot had already had his license withdrawn, and resigned from the airline and many observers saw no point in the prosecution being undertaken. He was fined £2000, but tragically took his own life shortly afterwards. (In this case no lives were lost, there were no injuries and there was no accident!)
- In 1991, a Korean pilot was sentenced to two years in prison after a fatal accident in Libya while attempting to land in bad weather. The copilot was given a suspended sentence of 18 months
- In 1993 at Paris, two pilots were indicted for homicide after a crash in which four passengers died, and findings of 'pilot error' were established at the inquiry
- In New Zealand, a pilot was prosecuted after a piece of cargo broke loose in flight and caused an accident

Representations made to ICAO by IFALPA, and to their own governments by its member associations, make it clear that pilots are not seeking immunity where the criminal law is broken, e.g. the smuggling of drugs or flying under the influence of drink or drugs etc. However, they are opposed to pilots being treated as criminals if they should suffer an accident as a result of human shortcomings such as an error of judgement or a shortfall of skill. It is emphasised by IFALPA that prosecutions under criminal law will have the effect of causing pilots to reduce, or perhaps end, their traditional willingness to freely provide information and evidence to accident investigators and to inquiries, if it becomes clear that evidence given under these circumstances at technical hearings may subsequently be used against them in criminal courts. This 'drying up' of safety information would be contrary to the public interest and the Federation is suggesting that evidence provided at technical hearings should be 'privileged' as follows:

'Evidence given at the technical investigation should be considered as privileged and not be available for use in any subsequent disciplinary, civil, administrative or criminal proceedings, nor for any public distribution.

'Evidence other than the final report should not be made available to any body which seeks to establish civil, criminal, or administrative responsibility or apply disciplinary measures. Nor should any of this information be made available for public distribution.

'Records used in the reporting of near accidents and incidents, voluntary or mandatory, anonymous or not, in a voluntary or mandatory system, shall not be made available for purposes other than accident prevention. In no case shall the identity of the persons involved be disclosed to the public.

'Any provisions relating to prosecution for violations of air regulations or criminal laws should be covered in legislation that is separate from aircraft accident investigation legislation and, where foreign pilots are involved, should encompass the principle of Transfer of Prosecution, where appropriate, to the state in which such pilots were licensed.'

It is probable that the objectives of IFALPA will not be achieved for many years, if ever, as most states are reluctant to consider the international consequences of any aspect of national criminal laws. Those states operating a Napoleonic Code of Law tend to treat all cases of death by accident in a similar fashion. Car drivers, train drivers, ships masters, hotel and apartment house owners are all subject to prosecution in the criminal courts should lives be lost as a result of an accident, whether caused by human or equipment failings, or a fire or structural collapse of buildings.

On 6 March 1987, Mr Engen, then Administrator of the FAA, made a speech to the 21st Annual Air Law Symposium providing support for airline pilots in their concern about criminal prosecutions of pilots (and air traffic controllers) involved in aircraft accidents. Mr Engen stated:

'An aviation accident reveals defects that we did not know existed. Unfortunately, in some cases people have resisted change, or new procedures until an accident demonstrated the need for change. Our procedures – which rely on systematic, comprehensive and open exchange of information with people involved in accidents – are essential to continue aviation safety progress...

'Both the FAA and the NTSB need to know as much as possible about an incident or accident to make timely and effective safety recommendations. In the best of all worlds, everyone involved in the aviation system, should speak openly and immediately to safety investigators, without fear that evidence acquired from a safety investigation will later be used in a criminal prosecution. Future aviation safety depends on that.

(Mr Engen commented on two cases in California where criminal prosecutions of a pilot and a controller were being contemplated by the legal authorities and then went on)...

'I am deeply concerned about the precedent that could be set here. Those people involved cooperated with safety investigators, even knowing that the FAA could impose a civil penalty on the pilot's certificate. I am not sure that the next pilot will cooperate, especially if he thinks that information provided to safety investigators might be used by other parties in a criminal prosecution. If such information were regularly available for use by prosecutors, I can easily understand why attorneys would have to advise clients against cooperating in safety investigations.

'We do not necessarily wish to exclude prosecutions under State criminal codes. To cite only one obvious example, most laws prohibiting flying under the influence of alcohol or drugs are State laws. They need strong enforcement, and we fully support local authorities' efforts to enforce them. Responsible local officials should recognise that effective enforcement of laws prohibiting drunk flying differs from most aviation accident investigations.

'Effective enforcement prosecutions must be conducted in ways that do not impair national air safety activities. The flying of an airplane is not easily governed under criminal procedures. The highly technical, skilled and instantaneous decisions that pilots must make in the cockpit should not, in normal cases, be second-guessed under the leisure of judicial proceedings.

'Safety demands that the pilot devote full attention to the proper operation of the aircraft, unimpaired by distractions that might arise from ill-conceived legal procedures. I doubt that any of you would feel comfortable as a passenger on an airplane whose pilot was more concerned about possible criminal charges than safe operations...

Mr Engen ended his speech by stating:

'We will continue to hold strongly to those precepts which provide for safety and air commerce. I share all legitimate concerns about air safety but more than legitimate concern is necessary for effective improvement.'

A different legal problem affecting airline pilots is the gap considered to exist as to pilot responsibility should an accident occur during an automatic landing in non-visual conditions, or a collision occur in air-space where little or no attempt is made by a State to provide air traffic control or communications. A possible conflict of interest exists between the State responsible for providing airports, air traffic services, or certificating an aircraft type, and the pilot should an accident occur and a solution to these legal problems is urgently required. This solution may be brought forward when accident investigations automatically take evidence from CEOs of regulatory authorities, manufacturers, airlines, airports and providers of air traffic services!

10: International Law, Security and Aviation Crime

Civil air transport suffers more from illegal acts than most industries, perhaps because of its international importance and high degree of visibility ensuring publicity for those who attack it for their own purposes. Illegal acts threatening the wellbeing and safety of the industry, are described by ICAO as 'Crimes of Unlawful Interference', and include hijacking and other acts of terrorism. The threat posed by these crimes in the 1960/70s brought about a flurry of international law-making in order to meet the threat, and civil aviation acquired a number of laws specific for its needs.

International laws and agreements

The international conventions and national laws dealing with aviation affect the legal status of the aircraft commander with regard to his or her responsibilities and authority to deal with happenings, and persons, that may affect the safety of a particular flight. The commander's basic position under the law is set out in two annexes to the Chicago Convention.

Annex 6, paragraph 4.5.1 states: 'the pilot-in-command shall be responsible for the operation and safety of the aeroplane and for the safety of all persons on board during flight-time.'

Annex 2, paragraph 2.4 states: 'the pilot-in-command of an aircraft shall have final authority as to the disposition of the aircraft while he is in command.'

Other international conventions

The conventions that deal with crimes of unlawful interference with civil aviation, give the pilot authority to deal with persons attempting to carry out, or who may attempt to carry out such crimes, as well as setting out other powers and responsibilities of the aircraft commander. As early as 1937, a convention on 'Prevention and Repression of Terrorism' was agreed, making acts of sabotage an international offence, and the 'Geneva High Seas Convention' of 1958 made brief mention of air piracy, but went no further than asking states to cooperate.

A dramatic increase in the number of attempted hijackings in the 1960s and 70s, resulted in national and international efforts to contain the problem. The Tokyo Convention of 1963 on 'Offences and Certain Other Acts Committed Aboard Aircraft' is regarded as the first useful legal instrument dealing with the crime of Unlawful Interference with Civil Aviation.

The Tokyo Convention of 1963

The Convention deals with the subject of jurisdiction over criminal offences and acts jeopardising the safety of an aircraft committed on board that aircraft while in flight, and it generally expresses that the State of Registry of the aircraft is to exercise jurisdiction. The only specific offence covered is that of unlawful seizure (hijacking), but the Convention gives the aircraft commander authority to restrain persons, and it makes provision for the aircraft, crew, passengers and cargo to be allowed to proceed as soon as practicable after the hijacking ends. There were only 16 contracting states in 1963, and the Convention did not receive the 12 Ratifications needed to come into force until six years later.

The Hague Convention of 1970

This Convention came into being as an attempt to suppress unlawful seizures of aircraft. It agrees on what should happen after capture to a person or accomplice who, on board an aircraft in flight, unlawfully by force or threat of force, or by any other form of intimidation, seizes or exercises control of the aircraft or even attempts to do so. The Convention permits jurisdiction by either the State of Registry or the State where the aircraft lands with the alleged offender still on board, and it requires that the offender be taken into custody and be extradited *or* considered for prosecution (i.e. a case must be presented to the prosecuting authority). The Convention also requires that unlawful seizure be added to the list of offences for which an alleged offender may be extradited. The Convention received 51 signatures and took only 10 months to obtain the necessary 10 ratifications to bring it into force.

The Montreal Convention of 1971

This Convention for the 'Suppression of Unlawful Acts against the Safety of Civil Aviation', deals additionally with acts performed from outside the aircraft. It contains similar provisions for punishment or extradition as the Hague Convention, but deals with terrorist actions other than hijacking, such as bombings and sabotage. The Montreal Convention secured 31 signatures and achieved the ten ratifications necessary to come into force in late January 1973.

These three Conventions are now the most widely accepted of all international conventions.

Diplomatic Conference, Rome 1973

ICAO convened a Diplomatic Conference in Rome in August 1973. It considered four proposals that set out to create an Enforcement Convention that would agree sanctions to be taken against states that did not abide by the other three Conventions which were binding only on those states which had signed and ratified them. The proposals did not attract the necessary support and the Conference ended in failure.

Informal agreements

Although the three International Conventions gained wide acceptance, hijackings continued and it was believed that weapons, false passports, money and most important of all, sanctuary, were being made available to hijackers by some states. In response to that problem, some other states include Crimes of Unlawful Interference and the coordination of anti-terrorist activities, as items for discussion when bilateral air transport agreements are negotiated, and the development may be regarded as a form of substitute for the failed Rome Diplomatic Conference.

A most effective agreement reached as a result of that development was between the USA and Cuba where the 'popularity' of Cuba fell from a high of 63 in 1969 to a low of 0 in 1976, although the effect of improved security at USA airports must also be taken into account.

The Bonn Declaration of 1978

The 'agreement' with the greatest potential for substituting for the failed Rome Diplomatic Conference is the Bonn Declaration of 1978. Meeting as major economic states, the leaders of Canada, France, West Germany, Italy, Japan, UK and USA issued this statement:

'their governments will intensify efforts to combat terrorism. In cases where a country refuses the extradition or prosecution of hijackers and/or does not return aircraft, the Heads of State or Governments will take immediate action to cease all flights to that country: at the same time their government will initiate action to halt all incoming flights from that country or from any airlines of the country concerned'.

As the seven signatory states perform 61% of the world's scheduled passenger flights as measured by passenger/kilometers, and manufacture most of the world's civil aircraft, they have a collective power to destroy the civil air transport industry of any 'offending' state. They can achieve this by suspending its air service agreements and taking additional measures to withhold deliveries of aircraft, engines and spare parts. The 'agreement' has only once been openly invoked and that was against Afghanistan, hardly a major threat to the safety of international civil aviation. It is possible however that threats have secretly been made to use this potent economic weapon, and it may have been a factor in bringing some hijackings to a satisfactory conclusion. IFALPA called for the 'agreement' to be invoked during the hijacking of a TWA aircraft to Beirut in 1985, when it was obvious that the Lebanese Government was unwilling or unable to prevent the hijackers from receiving supplies and reinforcements while the aircraft was on the ground at Beirut.

The Bonn Declaration overcomes the difficulty involved in obtaining near-unanimous agreement from more than 180 states attending major international conferences aimed at ending crimes of unlawful interference with civil aviation. It demonstrates that effective agreements are possible where the need to agree is accepted, and where states that block agreements can be excluded from the process. Some observers are disappointed that the provisions of the Bonn Declaration have not been used more frequently and effectively.

'Deals' with terrorists

There have been persistent rumours that some European governments and airlines made 'deals' with terrorist organisations and the states that allegedly give them support. In April 1986 the Los Angeles Times quoting unnamed USA State Department officials stated that both France and Italy made 'deals' with Libya in the 1970s so as to spare their citizens from attack, in return for terrorists sponsored by Colonel Gadaffi travelling freely through Europe. The same report alleged that the French Government made a similar 'deal' with the PLO, and that Greece and West Germany have maintained contacts on terrorism with Libya and the PLO. It is rumoured that Syria and Iran were involved in the destruction by bomb of Pan Am flight 103 at Lockerbie in December 1988 but were not coupled with Libya (in USA and British condemnation of the crime), because these two states were 'helpful' during the Gulf war against Iraq. It has also been alleged that Germany freed Palestinian terrorists already in their custody who were suspected of complicity in the same crime.

It seems probable that governmental contacts have been maintained with terrorist organisations and their protectors, and it is possible that some governments (and perhaps their national airlines) have attempted to obtain immunity from terrorist attacks. It is impossible to condone such initiatives, for the result would be to divert terrorists to other targets.

Further diplomatic initiatives

At the 26th Session of the ICAO Assembly in September/October 1986, a Resolution was adopted to develop a draft instrument that will extend the provisions of existing ICAO Conventions on unlawful seizure of aircraft and acts against the safety of civil aviation, to criminal acts at airports serving international civil aviation. These Conventions (Hague and Montreal) which have been ratified by 130 states, require the prosecution or extradition of offenders without exception, as well as the imposition of severe penalties. In 1989, ICAO, after considering the Lockerbie bombing, reiterated its policies that apply. In 1992, the United Nations Security Council adopted a series of mandatory sanctions against Libya, including a suspension of all flights to and from that country for its failure to 'provide a full and effective response to those requests so as to contribute to the elimination of international terrorism'. (The requests were for the extradition of two suspects in the Lockerbie bombing.)

AVIATION SECURITY SYSTEMS

International Civil Aviation Organisation (ICAO)

ICAO has a committee, with many states as members, to deal with Crimes of Unlawful Interference with Civil Aviation, but the committee's remit is mainly concerned with legal aspects of the problem. The technical aspects of security are assigned to an aviation security panel which meets regularly and as required by events with several states, IATA, IFALPA, Interpol and the ACI represented by specialists in security matters. Their task is to exchange information and to recommend minimum security standards to be applied by states in membership of ICAO. A separate group of experts meet to consider the problems of detecting explosives.

An important specification agreed in 1985 was for states to establish measures to ensure that operators (of aircraft) do not place or keep on board an aircraft, the baggage of any passenger who has registered but has not reported for embarkation, unless it has been subjected to security control procedures. Other suggestions put forward for consideration by states, included measures for better security control of transfer and transit passengers, measures to deny access to aircraft by unauthorised personnel and to prevent contact between passengers who have been screened and those who have not.

A further proposal requires states to implement measures to protect cargo, baggage, mail and operators' supplies and to establish procedures for inspecting aircraft believed likely to be the objective of unlawful interference. Although the measures proposed are obvious they would, if universally applied, greatly improve security at many airports where it hardly exists at all, particularly in the 'developing world'.

ICAO arranges regional aviation security seminars where states and organisations acknowledged to be in the forefront of the fast developing field of aviation security, provide advice and knowledge to less well informed states at no cost except that of being in attendance. An Aviation Security Manual has been produced and an effective system of training courses instituted, with 'developing' states in particular taking advantage of the courses.

International Air Transport Association (IATA)

IATA works in the field of security by operating a Security Committee comprised of its own experts and others from member airlines. The committee visits 'problem' airports and assesses the security measures in place before making confidential reports to the responsible authority. During the late 1980s, IATA publicised its airport security inspection program listing 24 inspections in 1986, 12 in 1988 and an expected 20 in 1990, but since then has kept silent about the program, which almost certainly continues. It made strong criticisms of the Athens airport incident in 1989 and publicised a plan to 'internationalise' responses to terrorism by:

- Forming an international advisory group to support governments when a hijacking occurs
- Setting up an international team of experts to investigate acts of unlawful interference after the event and recommend methods to prevent repetition
- Forming an international force to provide a military response to an incident if necessary
- Creating an international court to try captured hijackers/terrorists
- Establishing an international detention centre to hold sentenced terrorists

Some USA member airlines of IATA expressed concern at the time of the Lockerbie disaster that they were not allowed to use their own security experts at European airports where national agencies carry out that service, sometimes to a poor standard!

Interpol

The annual conference of Interpol has a standing agenda item to consider all aspects of international aviation crime. This includes hijacking and sabotage, with the greatest emphasis being placed on an exchange of information on the whereabouts of known groups of terrorists and the existing level and nature of the threat. ICAO, IATA, and IFALPA are represented by observers.

Other governmental agencies are involved in gathering intelligence about terrorists engaged in crimes of unlawful interference against civil aviation, and a major task is to continually assess the level of the threat and to apprehend the criminals involved. When a particular threat is identified, information is distributed on a 'need to know basis', and the authorities responsible then inform affected airlines and airport authorities on the level of alert that is required. Much of the information is highly secret and its distribution is affected by political considerations.

International Federation of Airline Pilots' Associations (IFALPA)

IFALPA has a standing committee that considers all aspects of aviation security. It is sometimes briefed by pilots who have themselves been hijacked, and is therefore fully aware of the dangers of hijacking and the need for a high standard of security at all airports exposed to the threat. One of the committee's main tasks is to develop policies for consideration at the Federation's annual conference, and to identify airports that fall short of the necessary standards of security, which are then considered as possible candidates for a ban to be imposed by the Federation's members.

In order to make possible a uniform method of assessment, the following items are considered in rating the achieved standards of security at airports:

- Existence of search and surveillance security of pre-boarding screening
- Existence of sterile areas beyond such screening
- Provision of specially trained personnel to conduct screening
- Provision of specially trained law enforcement officers
- Existence of airport security plan in regard to handling of aircraft subjected to unlawful seizure or bomb threat
- Facility for allocating isolated aircraft parking areas
- Security of airport operational areas including perimeter fences
- Security training programs for airport personnel
- Coordination between airport security authorities and appropriate Member Associations
- Provision of adequate bomb disposal facilities

Airports Council International (ACI)

ACI introduced airport passenger checked–baggage screening systems and plays a leading part in the planning, coordination and installation of new screening systems. It tracks developments aimed at strengthening measures to prevent cargo, courier and mail from being used to introduce explosive devices onto aircraft.

Although there is cooperation between all these organisations, they have each identified a particular role for themselves and together they form an effective security system, provided that states and airport authorities make sufficient resources available to follow the security recommendations made.

AIRPORT SECURITY SYSTEMS

Governments, airlines, security forces, airport authorities, pilots and passengers know that international laws will not prevent attempted crimes of unlawful interference and it is therefore necessary to provide security systems at airports.

These systems require special equipment and trained personnel, but the strength of public opinion since a wave of hijackings in the 1960s has forced authorities to accept the task and consequential costs. The hope was that effective use of international laws and good security at airports, would ensure that would-be criminals would be deterred by the difficulties of carrying out their intended crimes, and the certainty of punishment should they make the attempt. These hopes were doomed to failure by the ingenuity and fanaticism of the criminals, lack of commitment by some authorities, complicity of some states and poor performance of some security systems and their operatives and supervisors. Hijackings and bombings therefore continue to occur although airports with excellent security systems are almost immune to the problem.

A good airport security system has many of the following elements.

X–ray baggage screening

The latest equipments 'view' the item from all angles and can detect fine copper wire behind steel plate. Plastic explosives and bombs can also be detected by other equipment that uses two curtains of collimated X-rays operating at different frequencies and levels of intensity. The weaker scan shows plastics as solid shapes while the stronger one shows metals in black. Early systems had the problem that the high metal-penetrating voltage made plastic transparent, and plastic guns such as the Glock 17 were not detected. A second technique uses backscatter detection where a low-energy narrow-width X-ray beam scans a target. Attempts have been made to devise a system that is automatic, needing no video or human operator, but sounding an alarm when a suspicious image is found.

The efficiency of this component of the security system is only as good as the state of alertness of operators of the equipment, and it is essential to limit the period of time an operator works at a monitor screen to a maximum of two hours on any occasion. A new development in this field is an X-ray system built into airline check-in desks which can identify explosives and show them in colour on a monitor screen. A passenger's checked baggage is simultaneously weighed and X-rayed, and security staff observe the contents of the bags on remote television displays.

Suspicious items of baggage can then be held at the check-in desk, and the passenger questioned. Several USA airlines have bought the equipment in order to meet FAA requirements for the screening of baggage.

Metal detector gateways

Early models gave many false alarms, but sensitivities are now more accurately set and performance has been improved.

Body searches

This simple but effective technique slows the 'processing' of passengers and is therefore used only when metal detectors indicate a cause for suspicion.

Explosives detection systems (EDS)

New electronic devices with computer controlled logic, promise to detect explosives that are tightly wrapped in metal foil or cling film. They can detect minute quantities of emitted gases, and by use of a computer memory identify and indicate the substance. Even plastic explosives can be detected.

Thermal Neutron Analysis was seen as a possible system, but the machines weigh ten tons and cost one million dollars each. For this reason they are in use only at JFK New York, Dulles Washington, San Francisco California and Gatwick London, and it seems unlikely that more will be installed. Explosives trace detection devices have been demonstrated for the screening of electrical devices at La Guardia, Dulles and Atlanta airports.

FARs require air carriers to use EDS to scan checked baggage on international flights, and in the 1993 FAA annual report to Congress, it was stated that planning for certification testing of EDS is complete and system testing will be at the FAA Technical Center. The scope and timing of deployment will be phased in on a threat-assessment basis, with the first systems going into the highest threat location overseas. An integrated security system at Chicago O'Hare is an example to all, but it cost $40 million and took three years to implement, and it seems that on cost factors alone, the ICAO goal of a uniform standard of airport security is as far away as ever!

Sniffer dogs and gerbils

These specially trained animals are remarkably efficient at detecting explosives and drugs, and cost much less to operate than other equipment. The FAAs K-9 team program was established in 1972, and in 1993 had 32 participating organisations, comprised of airport operators and local police departments at 31 airports.

Psychological screening

Airline ticketing and check-in personnel receive training in this form of detection. A key element is the use of psychological profiles of typical hijackers, but organised groups of terrorists are aware of the technique, and make efforts to beat the system. They dress as businessmen and women, assume the identities of professional persons or some other guise, buy their tickets separately at different locations, and are not seen together until the hijack begins.

Separation – air/ground-sides

In order to reduce the problem of providing security at airports, security screening is applied only to those members of the public who are to travel, and it is therefore essential to keep the two categories of the public separate after the security screening process is carried out. The best standards of airport design make the process simple by controlling the flow of passengers down 'sterile' corridors and into lounges that are closed to the public, but in older airports problems continue to arise where the 'sterile' concept is difficult to apply.

Separation – transit/joining/arriving passengers

As some states reduce costs incurred in providing security by applying none, or a lower standard to passengers on domestic flights, or even provide active or passive assistance to groups of terrorists, it is necessary to prevent weapons or explosives being transferred from one flight to another at airports. Difficulties arise in achieving this objective because of the physical design characteristics of some airports, and airports where passengers have to transfer from one terminal building to another are a particular source of difficulty. The solution in these cases is to treat transfer passengers as joining passengers and apply security screening to them.

Identification – airline/airport employees

These persons should be subject to special security measures, particularly those who perform duties requiring them to pass from the 'ground' to the 'air' side. They are normally required to wear identity badges that include a recent photograph, but the system is only as good as the checks made on the identity of the employees. Employees who have the use of vehicles to deliver supplies from the 'ground' to the 'air' sides are a particular source of risk. There have been a number of hijackings and bombings where it is suspected that airport employees supplied terrorists with their deadly 'tools of trade' on board the aircraft, so protecting them from detection by airport security systems.

It is estimated that as many as 40 000 persons have access to the 'air' side of some major airports. It is hoped that identity cards that are electronically encoded and can be read at ranges of two to three feet by 'interrogators' will speed the flow of workers without reducing standards of security. With this system it would be possible to validate identity cards for particular areas of the airport if that were desirable. The electronically encoded cards have the further advantage of being readily cancelled if lost, stolen or withdrawn.

The need to apply security access control at USA airports arose when an armed former airline empoyee used an apparently valid ID card to gain access to an airliner, shot a passenger, the crew and then himself, causing a crash that lost 43 lives. It is estimated that resultant costs of implementing identification systems are one billion dollars for 274 major airports.

Security patrols

Major airports occupy many square miles of land, and it is impossible to provide perfect security. A great deal can be achieved by instituting a system of irregular but frequent patrols by trained security staff, and by inculcating a sense of security awareness among employees at the airport and the general public.

Screening of cargo

A change of tactics by would-be terrorists switching from hijacking to sabotage by bombs, results in a need to apply security screening to cargo, particularly when a high level of risk is present. Special equipment is used to expose cargo to the levels of pressure experienced in flight, for some explosive devices are operated by pressure switches. The enormous task of ensuring that postal items carried on board airplanes do not pose a threat to safety, is seen when it is considered that the USA postal system passed 2.5 billion lbs of domestic mail to American carriers in 1994.

Screening of duty-free goods

Duty-free goods should be subject to security screening, for hijackers have used flammable liquids as bombs to enforce their demands.

Unaccompanied and interline passenger baggage

Terrorists have transferred baggage between two or more airlines as a means of placing bombs on board aircraft, and it has proved necessary to devise new methods of checking this baggage and reconciling passengers and bags in order to prevent terrorists from placing bombs on board while they themselves are safe. In most countries an ICAO standard is enforced that requires bags and passengers to be 'reconciled'.

International versus domestic flights

There should be no difference between international and domestic flights in the security screening provided, because of an increasing tendency for domestic flights to be the prime target for terrorist actions. As more international flights are operated by wide-bodied aircraft, hijackers are tending to select narrow-bodied aircraft operating on domestic flights as targets, because hostage control is more difficult on the larger aircraft where there are more passengers and crew in a larger number of separate cabins, sometimes on two levels.

In September 1976, a TWA B727 was hijacked in New York and flown via refuelling stops to Paris. On 14 March 1977, random screening of flights was in progress at Barcelona airport, and a hijacker booked on flight IB205 to Palma found it was being security screened and changed his booking to the next flight IB027, which was not being screened. Boarding the aircraft carrying a handgun and a rifle, he hijacked the flight to Abidjan, Seville, Turin and Zurich.

Advanced security systems for aircraft

New technology may provide aircraft systems having the potential to help terminate hijackings by making information from aboard continuously available to security forces. It is proposed to install hidden video surveillance cameras relaying clear pictures of hijackers and their movements to security staff on the ground. Developers of one system state that their interest in producing the equipment was stimulated by the tragedy that befell the passengers of an Egyptair B737 that was hijacked to Malta in November 1985. Fifty-seven people died when Egyptian commandoes stormed the aircraft without first gaining knowledge of the disposition of terrorists, passengers and crew members.

The surveillance cameras are small enough to be hidden behind cabin furnishings, and to take pictures through tiny pinhead-sized holes that can be transmitted from the aircraft by HF radio and received on a TV screen at the local security center or by a video printer. The cameras provide clear pictures, even when an aircraft cabin is in darkness and the cost of equipping a B747 is stated to be only $90 000 and a B737 only $45 000.

Israel Airline El Al security

It is paradoxical that the airline most threatened by organised groups of terrorists is the one which best provides safety. El Al thwarts terrorism by applying its own very high standards of security. *All* baggage, including that intended for carriage in cargo compartments, is opened and inspected by highly trained security experts. Intending passengers are questioned on their reasons for travel and requested to provide confirmatory evidence to support statements made.

On 17 April 1986, a female passenger of Irish nationality who had already passed through London Heathrow Airport security checks, was found by El Al security personnel to be carrying ten pounds of detonator primed plastic explosive provided by her Arab lover, and timed to explode during an early stage of the flight. The bomb was hidden in the bottom of a carryall and the passenger had no knowledge of it, but an alert El Al security man detected a discrepancy in the weight of the bag and averted a major tragedy.

USA security systems

Each year the Administrator of the FAA is required to report to Congress on the effectiveness of the civil aviation security program. Highlights of the 1993 report are shown in Table 10.1.

Each year the FAA checks the effectiveness of screening at airports and the results are shown in Table 10.2.

Table 10.1 USA Civil Aviation Security Program.

	1989	1990	1991	1992	1993
Persons screened (millions)	1113.3	1145.1	1015.1	1110.8	1150.0
WEAPONS DETECTED					
Firearms	2879	2853	1919	2608	2798
Handguns	2397	2490	1597	2503	2707
Long guns	92	59	47	105	91
Explosive/incendiary devices	26	15	94	167	251
Other dangerous articles†	390	304	275	2341	3867
PERSONS ARRESTED					
Carriage of firearms/ explosives	1436	1336	893	1282	1354
Giving false information	83	18	28	13	31

† Since 1992, other dangerous articles include: stunning guns, chemical agents, martial arts equipment, knives, bludgeons and certain other designated items.

Table 10.2 Test object detection rate (%).

1987	1988	1989	1990	1991	1992	1993
79.5	88	92.3	93.8	94.5	94.8	94.5

Civil penalties totalling more than $1.2 million were proposed in 1993 against 27 airlines for failure to detect objects during FAA checks.

- There were no incidents of explosive devices detonating on board US or foreign carrier aircraft in 1993 and no hijackings recorded in the USA, or on board US registered aircraft, in 1992 or 1993
- During 1993, Federal Air Marshals provided in-flight security on selected flights of all major US carriers to and from 26 cities in 24 countries
- 18 security liaison officers are assigned to Europe, Africa, Latin America, and Asia
- 221 US air carriers are required to follow FAA approved security programs
- 204 foreign scheduled and charter carriers serving US airports are required to adopt and use FAA approved security systems
- These 425 carriers serve 428 airports within the US that have to use approved security systems
- Air carriers must use FAA approved explosive detection systems to screen checked baggage on international flights

- The FAA has begun a study on trade-offs between explosive detection levels, blast management, blast containment techniques, airplane structural enhancements and the costs and investment levels required
- The FAA's programs are based on the facts that terrorists do not normally strike at random and that the threat to any carrier is largely a function of its nationality. Counter-measures need to be high only when the threat is high
- The FAA performs scheduled on-site evaluations of foreign airports served by US carriers or each last point of departure for foreign carriers serving the USA. Two hundred and forty-four airports meet these criteria
- 126 evaluations were made in 1993, resulting in 534 security recommendations to foreign governments
- Lagos Nigeria failed to make the security improvements required after an evaluation and all services to and from the USA were suspended
- The FAA provides two security specialists to ICAO at no cost to the organisation
- The FAA trained 402 international students and assisted a number of foreign governments with security surveys and follow-up training needs

A cost-effective USA rewards program for terrorism information operated and funded by the State Department, Airline Pilots' Association and the Air Transport Association had by 1992 received information from 60 countries, and the largest single award was $1 million shared on a 2:1:1 basis. The chairman of the inter-agency rewards committee stated that during the Gulf War 'information received on an attack in South-East Asia allowed us to prevent an attack in which dozens, if not hundreds of lives would surely have been lost'. The program (and other intelligence) generates alerts that are passed by the FAA to affected airlines and airports. There are four levels of alert that require specific anti-terrorist measures to be implemented.

The continuing security threat

A worldwide threat to international civil aviation continued into 1994, and ICAO reported the following incidents for that year as shown in Table 10.3

In assessing trends in acts of unlawful interference, ICAO stated that the number of acts fell from 49 in 1942 to 39 in 1994, broken down into

Table 10.3 Unlawful interference with international civil aviation.

Category of occurrence	International	Domestic	Total
Sabotage	1	1	2
Seizure	10	13	23
Attempted seizure	2	2	4
In-flight attack	2	2	4
Facility attack	0	4	4
Others	0	2	2
Total	15	24	39

15 acts involving international flights, 20 involving domestic aviation and 4 attacks on ground facilities.

AVIATION CRIME

Criminals use civil aircraft including airliners to execute crimes and avoid the course of justice. Crimes such as hijacking and sabotage pose a clear threat to air safety, but others do not so obviously threaten the lives of innocent people.

Drug smuggling

Large scale smuggling of drugs is a major problem to the air transport industry, and a steady increase in trafficking and use prompted the UN to make a concerted global effort to control this scourge. The first international conference on Drug Abuse and Illicit Trafficking took place at Vienna in June 1987.

Commenting on problems posed to civil aviation by drug trafficking, an ICAO Bulletin stated that concealment of illicit drug shipments in the airframes of commercial aircraft necessarily involves the covert tampering and removal of panels and other components, and could affect air safety. Drugs are found in sections of aircraft to which only crew members and airline staff have access, including avionic bays, oxygen compartments and toilets. It is estimated that more than 80% of illicit drugs carried by air are flown on general aviation aircraft, and that smugglers frequently fly in an unsafe manner along unpublished and unpredictable routes, often at a low altitude as a means of avoiding detection. Forged pilot licenses are used to hire aircraft and these flights pose a serious threat to the safety of civil aviation.

The USA government has long known that Florida is a major gateway for the illegal importation of drugs, and takes extra precautions to defeat traffickers. Airlines that operate between Latin America and Miami are used to aircraft, passengers and crews being detained because of searches for drugs. In 1996, US Customs at Miami Airport adopted new methods by mingling undercover agents with passengers and looking for tell-tale signs of nervousness and guilt. Detection rates went up by 40%, three tons of drugs were seized in only six months and passengers were subject to few delays.

It is known that one of the most popular means of transporting quantities of narcotics in cargo, is to conceal them within frozen food-stuffs such as blocks of shrimps. Detecting and identifying narcotics concealed in this way is extremely difficult. In many countries it is feared that airline and airport employees with access to airliners are used to load/unload illicit cargo and that in some countries corrupt law and customs officers are also involved. Crimes of violence are common among drug dealers, and in 1989 it was suspected that drug dealers used a bomb to destroy a B727 near Bogota, Colombia in order to kill five police informants believed to be on board.

Most cocaine arrives in the USA aboard private aircraft which smugglers consider expendable, for even a twin-engined aircraft costs less than the profit made from sale of the cocaine it can carry. Some shipments are delivered by airdrop after which the pilot puts the aircraft on automatic pilot and bails out. In September 1985, a pilot laden with 79 pounds of cocaine was killed in Tennessee when his parachute failed to open.

In other incidents, drug smugglers have set out to evade US customs by covertly and illegally landing aircraft at secret or private airfields, or at public airports at night and without making contact with any agency. It is alleged that landings are made in total darkness and radio silence at airports being used by legitimate air traffic with a consequential risk of collisions in the air or on the ground. Other countries almost certainly have similar problems.

Drug abuse in civil aviation

The ready availability of drugs is a serious threat to all of human society, and particularly to the air transport industry which is especially vulnerable to mistakes made by pilots, controllers or mechanics while under the influence of drugs, and the NTSB attributed a 1983 fatal accident to illegal drug abuse. Two crewmen died when a cargo flight crash-landed at Newark Airport and autopsies showed that the pilot had been smoking marijuana, possibly in flight. In March 1985, a New York based air traffic controller who had been injecting three grams of cocaine daily at work, put a DC10 aircraft on a collision course with a private plane. A collision was narrowly averted when the private plane made an emergency landing. In another incident that did not affect safety, an airline computer operator who was 'high' on marijuana failed to load a crucial tape into the reservations system. The system was out of action for eight hours costing the airline about $19 million!

In 1990, Congress and the FAA decided to introduce random drug testing for all civil air transport employees who perform safety-related duties – pilots, controllers and mechanics etc. There have been few detected instances of drug abuse by airline pilots, and random testing provoked angry responses from USA ALPA that believes the practice to be an unjustified abuse of civil liberties. The extremely few cases detected by the checks (only two positive flight crew member tests out of 30 732 random tests conducted in the first six months, a rate of 0.007) led to a reduction in the original 50% testing carried out, but ALPA has not yet succeeded in having the number of tests reduced to its suggested 10%. ALPA argued that 50% random testing was wasteful at a cost of $24 m per year and that a 10% rate would be an ample deterrent. It pointed out that when the FAA reduced testing of its own employees from 50% to 25%, it had no impact on the effectiveness of the program. ALPA does *not* oppose drug testing for applicants for aviation safety critical jobs – pilots, air traffic controllers, aircraft dispatchers, aircraft maintenance and ground security employees etc.

Suspect/unapproved airplane parts

As early as 1957 the FSF was issuing warnings that the 'stream of parts necessary for the maintenance and overhaul of aircraft and engines has become polluted'. This safety problem continues to grow and accidents have been attributed to the use of bogus, damaged, time and service expired parts. In September 1989, a chartered twin-engined Convair turboprop aircraft with 50 passengers and five crew disappeared from radar as it approached Denmark. Investigation showed that bogus bolts used in the tail vertical stabiliser had failed, and the airplane had crashed with the loss of all on board.

In 1988 Boeing investigated allegations that it had been supplied with 100 000 counterfeit parts (ball bearings) which were then installed in airliners and military airplanes, and the supplier was prosecuted. In 1991 a United Airlines mechanic discovered a suspect engine bearing seal spacer and returned it to Pratt and Whitney. They tested the $500 ring and found that it was one-third as hard as it should be and would have failed after about 600 hours of use versus the 6000 hours of a genuine part. Engine manufacturer Pratt and Whitney has stated that it receives many calls from mechanics anxious to verify authenticity of parts being used, and that it sends warning circulars to alert customers. Its trademarks and packaging have been copied by unauthorised persons.

A Miami based aircraft parts supplier lost more than $1 million of its parts inventory to theft in 1992. In 1995, the FAA proposed new rules after a task force reported on accidents/incidents caused by counterfeit components and repairs. More than 250 infringements of the law were found and 100 successful prosecutions were made. Among measures proposed were increased cooperation with the Department of Defense, which sells surplus parts, particularly helicopter parts, to the commercial market, often with inadequate documentation for civil purposes. The FAA reported that it was cooperating with the Department of Transportation (DoT), FBI and police forces in pursuing active programs aimed at eradicating the problem, with prime reliance placed on FAA certification of parts manufacture and dealerships.

The DoT reports that after giving bogus parts a high priority, the number found increased from nine in 1990 to 52 in 1991, 362 in 1992, to 411 in 1994, 317 in 1995 and 230 in the first six months of 1996. When DoT inspected some FAA approved facilities in Europe, 45% of new parts and 95% of old parts were found to be improperly documented.

More than 100 000 brake parts illegally stamped with the same serial part numbers as those by an approved US supplier, were shipped to a Florida parts dealer by a German manufacturer. In the UK a major problem is the absence of regulation of parts dealers, and some proven cases involve forgery of FAA documentation and falsification of the 'safe life' remaining of parts.

A British Airways B747 caught on the ground at Kuwait when the war with Iraq began, and badly damaged in the fighting, was to have been destroyed by agents of the insurers after the war. However, many parts from this airplane appeared in used part catologs, and in one case were inadvertently bought by Virgin Airlines. The FAA raided 34 places in the USA to recover some of these parts! In December 1995, an American Airlines B757 crashed near Cali, Colombia and after initial investigation at the crash site, it was agreed to remove the parts to a compound near Cali Airport. A large number of parts were stolen and appeared on the spare parts market. The 523 parts 'lost' included two Rolls Royce engines and the digital flight data recorder.

Miami is believed to be the centre of trade in bogus parts, sometimes from damaged airplanes and sometimes manufactured from inferior materials or designs with false documentation provided. At nearby Fort Lauderdale, a pilot unable to complete his pre-flight checks found that two flight computers worth more that $250 000 had been stolen from his UAL jet to be sold on the thriving black market for stolen and counterfeit parts. Automotive parts have been used for airplanes or genuine parts improperly repaired, even domestic bed springs have been used in safety-critical aircraft applications. Some bogus parts have originated in China or Taiwan. One entrepreneur made 200 counterfeit parts and then ordered 200 genuine parts from the original equipment manufacturer. He swapped the parts and returned the counterfeit parts to the manufacturer, paying only a cancellation charge and then sold the genuine parts. The safety problem posed by use of bogus parts is likely to remain for many years.

Hijacking

The crime of hijacking has been perpetrated over a period of about 60 years, and it is generally agreed that there are three main motives for the crimes. They are committed with the intention of securing political or criminal objectives, and less frequently by persons of unsound mind for no practical purpose whatsoever. The earliest instance on record, occurred in South America in 1931 when a Ford Trimotor aircraft, operated by the USA airline PANAGRA, was hijacked by Peruvian revolutionaries who forced Captain Byron Rickard to fly to Chile. Captain Rickard was doubly unfortunate in again being hijacked thirty years later while flying a B707 for Continental Air Lines.

The earliest recorded case of a hijacking for purely criminal purposes was on 16 July 1948, when three Chinese villagers used pistols to take over a Catalina amphibious airplane operated by Cathay Pacific Airways on behalf of Macau Air Transport. The intention of the hijackers was to force the aircraft to land on the sea near their village and then to rob the passengers and steal the cargo. During the hijacking the pilot of the aircraft was shot and killed, and the aircraft crashed and was destroyed. All passengers and crew members were killed and only one of the hijackers survived out of the total number of 27 persons who were on board the aircraft.

The best known case of hijacking for criminal gain happened in the USA on 24 November 1971, when a white male, D.B. Cooper, hijacked a Boeing 727 by claiming to have a briefcase full of dynamite which he threatened to explode. After landing to take aboard five parachutes and a $200 000 ransom in $20 banknotes in exchange for the release of passengers, the aircraft took off and flew at 10 000 feet and 200 miles per hour over a remote area of forest and mountains in the northwestern States of the USA. Cooper ordered the crew to open the exit door in the tail of the aircraft and parachuted to the ground with the ransom money which was in a laundry bag. Cooper was never found, but as the weather was very cold and he was wearing only a normal suit, it is possible that he died after landing in the rugged mountains and forests of Washington or Oregon, or in the cold waters of the Columbia River.

Nine years after the hijacking, a boy found $6000 in water stained bills that were identified as part of the ransom money. Nothing more was found, although private and insurance investigators continued to work on the case, and in 1986 they were still engaged in dragging the Columbia River with the case still remaining open in police files. The crime was very carefully planned even to the making of fictitious reservations for a large number of passengers, causing the aircraft to depart with only one-third of the seats occupied. So many attempts were made to emulate Cooper's methods in that crime that the rear exit doors of the B727 and other aircraft types were modified so that they could not be opened in flight.

Among reported hijackings by persons of unsound mind, one particular case in May 1981 was that of a religious fanatic who hijacked an Aer Lingus Boeing 737 aircraft to Le Touquet, France where the aircraft was stormed by security forces without loss of life. In a similar case of an Air Inter aircraft at Orly, Paris in 1977, tragedy resulted when police stormed the aircraft from the rear and in the resulting confusion the solitary hijacker dropped his hand grenade killing one person and injuring three others. It is possible that the decision to storm the aircraft was wrong, but it is only with pre-knowledge of the hijacker's motives, identity, strengths, weaknesses and intentions, or with hindsight, that such mistakes can be avoided.

There is at least one recorded case of a 'hoax' hijacking when the cabin crew of a TWA aircraft on a New York – Geneva flight on 25 August 1978, found a note that appeared to be from a hijacker stating that a bomb was on board. The captain and crew accepted the note as genuine and after the aircraft landed in Switzerland as scheduled, all passengers and crew stayed in their seats waiting for the hijacker to make known his demands and reveal his identity. Nothing happened and after a period of time it became obvious that a hoax had been perpetrated.

Apart from hijackings committed for criminal gain or by persons who are mentally ill, the great majority are committed for political purposes with many taking place at times of tension for states and political organisations. Wars and political struggles between India/Pakistan, Iran/Iraq, Ethiopia/Somalia, China/Taiwan, the former two Germanys, USA/Cuba, Russia/Turkey, Russia/former states of the USSR, Israel/Arab states and organisations and China/South Korea have produced surges of hijackings, with hijackers having a mixture of objectives such as escape from political repression, drawing international attention to their 'cause' or obtaining the release from custody of fellow terrorists.

A perfect example of a link between political instability and hijackings was in 1990 during the Gorbachev regime in the USSR when there were 32 attempted hijackings compared with only three in the preceding year. All but one of the attempts were on Aeroflot airplanes with nine resulting in landing at foreign airports. During the same year 130 000 people were detained while attempting to carry prohibited objects or substances onto aircraft. An amazing 535 firearms, 51 867 knives and sidearms, about 100 hand grenades, 78 artillery shells, 500 kg of explosives and 'tens of tons' of poisonous and highly flammable substances were confiscated by the KGB, according to Anatoly Bondarev the then head of flight security at the Ministry of Civil Aviation.

A serious threat to safety

A huge upsurge in hijacking attempts in the late 1960s/early 1970s was seen by airlines and flight crews as a serious threat to the existence of civil aviation, because of the probable response of passengers to an increased risk of death or injury. In addition to risk arising from the willingness of hijackers to use their weapons of guns or bombs on the ground or, more dangerously in flight, it was considered likely that on some future occasion an aircraft would be involved in an accident because hijackers would insist in continuing a flight in unsafe conditions. In a number of early hijackings, hijackers refused to believe the captain when he stated that the aircraft had insufficient fuel to carry out the flight as they wished or that it was not safe to land at the selected aerodrome. As a result, there have been occasions when the lives of passengers were put at greater risk by the captain not being fully in command of the aircraft, than by violence threatened by the hijackers.

In some crimes, members of flight crew are killed, with a consequential increase in the level of risk experienced while in flight. In almost all cases flight crews are deprived of sleep for long periods of time and are subjected to abnormal levels of stress arising from concern for their own safety and that of their passengers. During long hijacks, aircraft receive no maintenance or servicing and crews operate aircraft to unfamiliar airports without receiving normal assistance from air traffic controllers and weather forecasters. Some states determined to avoid involvement in hijackings, close their airspace and aerodromes to hijacked aircraft in flight even when it is known that there is very little fuel remaining. All these circumstances are reason to end crimes of unlawful interference as soon as possible (see Fig. 10.1).

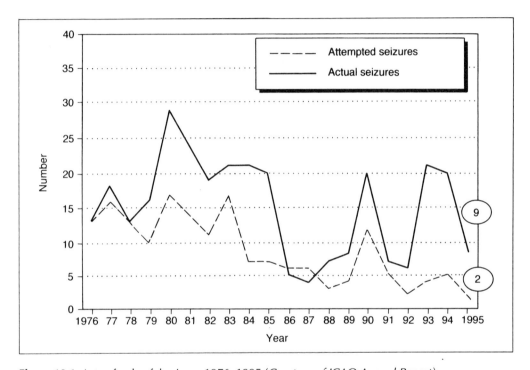

Figure 10.1 Acts of unlawful seizure 1976–1995 (*Courtesy of ICAO Annual Report*).

Unusual hijackings

Some hijackings have a particular facet that distinguishes them from all others, and these differences may arise from the objectives of the hijackers, the number of lives lost, the airports involved, the duration of the incident or any number of other unusual features.

Hijacking by mercenaries

In November 1981, the attempt of a group of mercenaries to overthrow the government of the Seychelles failed and they hijacked an Air India

B707 at Mahe airport in order to escape to South Africa. On arrival at Durban, government authorities released the hijackers before the passengers, and not until international pressures were exerted by IFALPA and governments signatory to the Bonn Declaration, did the government of South Africa arrest and put the hijackers on trial. It is widely believed that the government was implicated in the attempted coup, for many of the 'mercenaries' were members of the reserve forces of South Africa. It is also believed that this was the first occasion the signatories of the Bonn Declaration threatened to use sanctions against a state. All hijackers (except one) including leader 'Mad Mike' Hoare were found guilty of the crime of hijacking and sentenced to terms of imprisonment. Hoare was sentenced to ten years but released after three in view of his age (65) and it seems that a person young enough to lead a team of mercenaries may be too old to serve a sentence for the crime!

Triple hijacking in Venezeula, 1981

On 7 December 1981, three aircraft were hijacked from the Simon Bolivar Airport at Caracas, Venezuela, within a period of only 49 minutes, providing a remarkable comment on the level of security at that airport. One of the aircraft was a B727 belonging to Avensa and the other two were DC9 aircraft belonging to Linea Aeropostal Venezolana (LAV). All three flights were 'domestic' and a total of 325 passengers and crew were on board.

The hijackers were armed with sub-machine guns, pistols and hand grenades, and all the aircraft were flown to Barranquilla, Colombia, although one of the DC9s went there by way of Aruba in the Netherlands Antilles, where 21 passengers were released in exchange for fuel. At Barranquilla the hijackers demanded flight charts for all of Central America, fuel, food, water and medicine, with ten million USA dollars, release of a number of prisoners in Venezuela and publication of a political manifesto citing their political complaints and objectives comprising their main demands. Colombian authorities blocked the runways and refused to provide fuel and other supplies until the women, children and infirm male passengers were released.

The three aircraft left Barranquilla after about eight hours on the ground within a period of only a few minutes, and without their main demands being met. The B727 was flown to Guatemala City and the DC9s went to Tegucigalpa, Honduras, although one of them made a mysterious ten minute stop at an air force base in El Salvador while enroute to Honduras. More than forty passengers were released in return for fuel and all three aircraft then flew to Panama City. Negotiations with the Ambassador of Venezuela followed, and two more passengers were released before all three aircraft left for their final destination of Havana, Cuba. Here the 11 hijackers were taken into custody by the Cuban authorities, and the aircraft, passengers and crew were allowed to return to Venezuela.

Alitalia B747 – Delhi to Bangkok, June 1982

An unusual feature of this hijack is that a ransom was paid, and the hijacker permitted a period of freedom before legal action was taken against him. The aircraft was hijacked by a Sri Lankan passenger named Sepala Ekanayake who boarded the flight at New Delhi. After two hours of flight, he commandeered the aircraft claiming that dynamite was wired to his body and that he had three accomplices. The aircraft continued to its scheduled landing at Bangkok, arriving at 0315 local time on 30 June.

After arrival, the hijacker conducted negotiations with the Italian Ambassador to Thailand and a Thai Minister of Government. He demanded a ransom of 300 000 USA dollars, to be reunited with his Italian wife and son who were to be flown from Italy to Bangkok, assurance of safe passage to Sri Lanka and freedom for at least two weeks. During negotiations three passengers were released and then a further 98, with the remaining passengers being released when the authorities agreed to the demands of the hijacker. It was then found that the 'dynamite' was plastic tubes and that the three accomplices did not exist. Ekanayake collected the 300 000 USA dollars, was joined by his wife and child and permitted to leave Bangkok for Colombo aboard an airliner operated by Air Lanka. By 7 July, Ekanayake had been arrested and the Sri Lankan Parliament passed legislation to implement the three international conventions.

A marathon hijacking, TWA B727 – Athens to Beirut, 14 June 1985

TWA Flight 847 was routine until soon after take-off when a scuffle was heard behind the cockpit bulkhead. Captain Testrake ignored repeated banging on the flight deck door until Uli Derickson, the Flight Purser (senior stewardess), called using the intercom to report that the stewardesses were being 'roughed up', and asked 'Please open the door'. As the cockpit door was opened, two men dashed in, one carrying a large bore automatic pistol and plastic bomb, and the other carrying hand grenades with drawn safety pins. Captain Testrake reported later 'They were wild and demanded that we fly to Algiers'.

For the next sixteen days the eyes of the world were focussed on dangers faced by the crew and passengers of TWA Flight 847 with USA television networks using live news transmissions by satellite channels on a twenty-four hour basis.

During this long period, the airplane flew between Beirut and Algiers several times, sometimes taking off overweight and on other occasions landing with little fuel remaining. Runways were obstructed, the airplane had no maintenance, a US serviceman passenger was brutally murdered, the flight crew had little rest and the hijackers received assistance (including arms and food and replacement terrorists) while on the ground in Beirut. All TWA crew members behaved heroically, and it was due to them and strong diplomatic pressures exerted by the US and its allies, that only one life was lost – a US Navy diver.

Consequences of this hijacking

The inability of the Lebanese authorities to establish control over Beirut airport led to the hijackers being resupplied and relieved during the long period of hijacking and hostage taking. This caused IFALPA to apply a worldwide ban on operations to Beirut because of the total absence of security. IFALPA also joined with IATA in applying pressure to the Greek authorities, calling for early improvements to the security systems at Athens airport where the hijackers had boarded the TWA aircraft with their weapons. These pressures obtained quick results and security recommendations made by IATA were accepted by the Greek authorities.

Pilot organisations took another important step for the first time, when they declared that the absence of security at Beirut airport required every commercial flight that originated from that airport or transitted it, to be 'security quarantined' at its next point of landing. IFALPA required aircraft, cargo, baggage, supplies, passengers and crew members to be subjected to stringent security measures before being permitted to continue to another destination.

The FAA amended Federal Regulations dealing with airplane operator security within a month of the hijacking, and improved flight security coordination and flight crew training, and Congress acted decisively to enact additional legislation.

On 8 August, President Reagan signed the Foreign Assistance Act of 1985 which included provisions designed to strengthen methods of combatting air piracy both in the USA and abroad. The principal aim of the Act is to pressure foreign governments into improving flight security at foreign airports serving the USA. The Secretary of Transportation is required to assess security at various foreign airports and if airports fail to meet minimum security standards of ICAO, the Secretary is to notify the appropriate foreign authorities with recommendations for improvement. If foreign authorities do not make recommended improvements within 90 days, the Secretary must post warnings at all USA airports and inform all prospective passengers via the Federal Register, and in notices placed in airline tickets, of deficiencies at these foreign airports. He is authorised, subject to the approval of the Secretary of State, to prohibit any USA or foreign airline from offering service between a security-deficient foreign airport and the USA. In Frankfurt in 1989, Lebanese terrorist Mohammed Ali Hammadi was sentenced to life imprisonment for his part in this hijacking and the murder of the US Navy diver.

A pointless hijacking

In April 1990, an American Airlines aircraft with no passengers on board was seized by a lone gunman in Port au Prince, Haiti and held for several days. The gunman was a member of the Haitian airport security contingent and demanded to be flown to the USA. He eventually left the aircraft and disappeared into the darkness!

Hijacks of Ethiopian aircraft

In 1992, five Ethiopian Airlines airplanes on domestic flights were hijacked reflecting the twin factors of a poor security system and a disturbed political situation.

Intervention by security forces and in-flight resistance

Intervention by armed forces is the last resort of security, occurring only after hijackers have succeeded in taking full control of an aircraft and have refused to release the aircraft, passengers and crew. IFALPA policy calls for no intervention to take place unless specifically requested by the pilot-in-command, but there are occasions when the pilot is not able to make known his or her wishes and the policy accepts that governments principally concerned may consider that extraordinary measures to save human life are required.

Decisions to end hijackings by force have cost lives on a number of occasions and a major disagreement arose between the USA authorities and ALPA, the national pilot association, when, in November 1972 at Orlando Florida, security forces (FBI) shot out some of the tires of a Southern Airways DC9 aircraft as it was about to take off for Cuba for the second time in that hijacking. The hijackers forced Captain Haas to continue the take-off and it is regarded as a near miracle that the aircraft was not destroyed then or during the subsequent landing, as a direct result of the actions of the security forces.

Captain Haas' representations and a review of the incident caused better procedures to be introduced in the USA to ensure that the safety of passengers and crew has priority over attempts to apprehend hijackers.

There have been 'shoot-outs' in flight between armed guards and hijackers. In the autumn of 1970 when an El Al aircraft was being hijacked in British airspace by a Palestinian group that included the notorious Leila Khaled, live hand grenades were rolling around the passenger cabin while a struggle took place in the aircraft. The hijackers were overpowered and taken into custody by British authorities after the aircraft landed, but were released without serving long terms of imprisonment to the everlasting shame of the British Government and Prime Minister Edward Heath, who gave way to threats of reprisals from the PLO.

Armed guards aboard an aircraft belonging to an African airline are reported to have beheaded a would-be hijacker in full view of the passengers.

In March 1985, a lone hijacker entered the cockpit of an in flight Saudi B737 and demanded to be taken to Teheran, Iran, but was persuaded by Captain Nagadi to allow the airplane to land at Dhahran for fuel. During negotiations with the hijacker at Dhahran, Nagadi managed to evacuate passengers and cabin crew, but when security forces boarded the airplane the hijacker dropped his bomb and it exploded under Nagadi's seat. The copilot was uninjured, the hijacker was killed and Nagadi received serious injuries including the loss of one eye. For his courage, Nagadi received the FSF's President's Citation for Professionalism and Valor. In the same year, a Colombian presidential candidate was shot and killed by a lone assassin aboard an Avianca domestic flight in Colombia.

In view of all these circumstances, it is to the great credit of flight crews, and particularly captains, that there were few instances of loss of life occurring by accident to the aircraft during early hijackings. In view of the inherently hazardous nature of any flight conducted in abnormal circumstances, it is inevitable that such accidents will occur if hijackings continue to take place (see Figure 10.2).

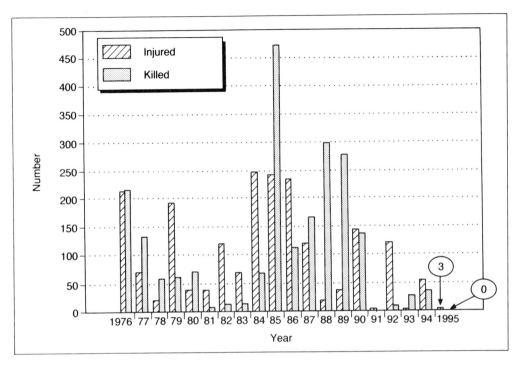

Figure 10.2 Number of persons killed or injured, 1976–1995 (*Courtesy of ICAO Annual Report*).

A hijacking that ended in disaster

On 25 December 1986, an Iraqi Airways Boeing 737 took off from Baghdad for Amman, Jordan, with 107 passengers and crew on board. The flight was scheduled to take only one and a half hours and had been in progress for 60 minutes with the passengers taking their lunch, when one of a group of four would-be hijackers ran toward the cockpit. He was intercepted by one of six plain clothes Iraqi security men on board, and a bloody exchange of shots began. At a later stage of the battle, two hand grenades exploded, one in the cockpit and one in the rear cabin. Although the aircraft was severely damaged, the injured pilot managed to fly towards an emergency aerodrome at Arar in Saudi Arabia, in an attempt to make a landing. Seventeen minutes after the start of the attempted hijacking, the B737 crashed, broke up and caught fire, just before reaching Arar. Sixty-two of the persons on board were killed and 20 more were injured.

It is highly probable that other incidents of this type will arise when there are in-flight 'shoot outs' between hijackers and security forces, or when the circumstances of a hijacking result in the flight crew being reduced in competence by reason of fatigue or injury.

Armed interventions

Among successful examples of armed intervention must be listed the Israeli rescue of crew and passengers of the Air France Airbus A300 at Entebbe in 1976, and the West German rescue of a hijacked Lufthansa B737 at Mogadishu, Somalia in October 1977. This second event occurred after the hijackers had shot and killed the Captain of the B737 when the aircraft was at Aden before flying to Mogadishu.

A disastrous attempt at intervention occurred at Malta on 24 November 1985, when a force of Egyptian commandoes made an attempt to rescue the passengers and crew of a hijacked Egyptair B737.

Flight MS 648 had departed Athens for Cairo on 23 November under the command of Captain Hani Galal with a total of six crew members and 89 passengers on board. Five of these passengers were hijackers, all Arabs, one with a Tunisian passport and two with Moroccan passports. The hijacking took place soon after take-off, and Captain Galal was ordered to fly to Malta. The hijackers were armed with revolvers and hand grenades, and it is believed that one of them gained access to the weapons after passing through the normal security checks and while identifying his baggage during an extra security check at the aircraft. While en-route, there was an exchange of fire between the hijackers and one of the four Egyptian security guards, resulting in a hijacker being killed and the security guard being seriously wounded. After the in-flight shooting, the captain initiated an emergency descent from 35 000 feet to 14 000 feet in order to reduce the risk of the cabin differential pressure rupturing the damaged fuselage.

After initial refusal, Maltese authorities gave permission for the B737 to land at Luqa airport after the captain reported that the aircraft was short of fuel because of flying for two hours at a relatively low altitude following the emergency descent. The aircraft was on the ground at Malta for about 22 hours, during which time the hijackers began to carry out their threats, killing four passengers to put pressure on the authorities to refuel the aircraft. Once the killings began, the Prime Minister of Malta decided that the aircraft would *not* be refuelled and it was inevitable that more blood would be spilled before the incident was over.

On the morning of the 24, food was supplied to the aircraft and shortly afterwards 11 women passengers, two injured stewardesses and the wounded security guard were released, after one of the hijackers had again shot him, stating, 'instead of having him injured, take him dead'. An American woman passenger was also shot at short range and then thrown down the aircraft steps; somewhat miraculously she was not seriously injured.

Captain Galal became convinced that the hijackers would continue to kill innocent passengers and succeeded in passing a message requesting that an armed intervention be mounted to free the hostages. The Maltese authorities did not have suitable forces of their own, and are reported to have refused to allow a highly trained unit of the USA armed forces to intervene, or for a USA military aircraft carrying special equipment to land at Malta. That equipment is reported to have been transferred to an Italian civil aircraft at an Italian airport, and to have been on its way to Malta when a unit of Egyptian commandoes carried out a rescue attempt that resulted in heavy loss of life.

The attempt is reported to have taken place without any coordination between the affected parties, and while the airport's fire and rescue services were changing shifts. An absence of sophisticated anti-terrorist equipment meant that the intending rescuers did not have information on the precise location of the passengers, crew and hijackers on board the aircraft. Because the hijackers were believed to have secured the main aircraft entrance doors from being opened from the outside, a force of approximately 25 Egyptian commandoes used explosives placed in the rear cargo-hold to blow a hole in the cabin floor to gain access. As soon as the hijackers realised what was being attempted they added to the casualties and confusion by exploding three hand grenades.

A serious fire broke out in the passenger cabin and the gun fight that followed took place in the worst of conditions with a large number of casualties being the result. The captain attacked one of the hijackers by hitting him with a fire axe and was himself injured in the struggle that followed. Only 25 people escaped injury and all but one of the hijackers were killed. The full casualty list reads:

- 60 fatalities including two stewardesses killed during the intervention
- 22 passengers injured before or during the intervention
- 2 stewardesses injured by hijackers' bullets and the copilot also injured

Some casualties were caused by fire and smoke and the airport fire brigade did not succeed in reducing the number of casualties as they might have done had they had been pre-warned of the circumstances of the rescue attempt.

The sorry events of Malta caused IFALPA and others involved in aviation security, to take the view that armed Sky Marshals are inappropriate in most circumstances although it is clear that states that believe themselves to be at great risk, such as Israel, will continue to carry armed guards aboard their aircraft. It is to be hoped that the lessons learned at Malta will discourage armed interventions except as a last resort, and when adequate arrangements have been made with highly trained forces equipped with the latest technology being used.

Other acts of terrorism

Terrorists find hijacking is becoming difficult to achieve, and increasingly turn their attention to other forms of terrorism that may meet their objectives of putting pressure on Governments and bringing their aims and beliefs before the general public. Options include the use of missiles, bombs and guns against aircraft, airport installations and airline premises, and all have been used within the past ten years and will probably be used again. Effective counter measures are difficult to apply because of the international nature of terrorism and an increasing tendency for the various groups to cooperate together.

Missile attacks

In 1986/7, the Mujahedeen of Afghanistan were using surface to air missiles, obtained from British and USA sources, in Afghanistan and shot down a number of aircraft, some of which were reported to be aircraft of Ariana Airlines carrying civil passengers. There is little doubt that the UK and USA governments supplied the weapons or knew of their supply. It is believed that many hundreds of portable hand-held surface to air missiles were sold by the original Afghan recipients and transferred by various routes to the former Yugoslavia for use in the fratricidal wars of the mid 1990s. It is a matter for concern that some of these weapons may be obtained by other terrorist groups and be used as substitutes for hijackings and bombing.

Less successful missile attacks have been launched against aircraft in flight and on the ground at Paris, Rome and Beirut airports, and it is probable that these attacks will be repeated if the terrorists can obtain missiles. Should this happen, it will be difficult to prevent their use against aircraft flying pre-determined flight paths in the vicinity of airports, and against airport installations, for the weapons are easy to transport and simple to use. A deterrent against that possibility is that missiles are difficult to smuggle across frontiers and it is relatively easy to trace the source of supply.

Armed attacks on airports and airline premises

The very large number of possible targets available to terrorists engaged in this type of activity make it almost impossible to provide effective security. The best safeguard available, is a high level of awareness of the threat by the general public and airline employees, plus an effective use of counter intelligence.

- One of the first of this type of attack was a suicide attack by three Japanese Red Army terrorists at the Ben-Gurion airport at Tel Aviv in June 1972, when they used automatic weapons and grenades retrieved from checked baggage to indiscriminately kill 27 persons in the arrival hall.
- In August 1973, Arab gunmen killed three people and wounded 55 in an attack on Athens airport, and in December 1973 another Arab terror squad blasted a Boeing 707, seized another plane and fired machine guns at passengers at the start of a hijack at Rome airport.
- In June 1985, a Boeing 727 belonging to the Jordanian airline Alia, was blown up at Beirut airport after being seized by Shiite Moslems.
- In December 1985 simultaneous attacks using grenades and guns were made on the Rome and Vienna airports, causing the loss of 20 lives and leaving scores injured.
- On 9 and 11 March 1994, two terrorist mortar attacks were made on London Heathrow Airport, when mortar shells were fired from an airport hotel car park and aimed at the nearby runways. No serious damage was caused but security authorities were alarmed at what was seen as a growing problem, for there were only 10 such incidents causing 9 deaths and 123 injuries in 1992, but 29 with 27 deaths and 2 injuries in 1993.
- On 20 July 1996, a bomb at Reus airport, NE Spain caused 36 casualties.

Bomb attacks on aircraft

One of the earliest known incidents of this type happened to a Viking aircraft of British European Airways on 13 April 1950, while over the English Channel on a flight from London Gatwick Airport, to Paris Le Bourget, when a bomb exploded in the rear toilet compartment (see Figure 10.3). There was a loud noise, a sudden surge of pressure, and the cockpit to cabin door was blown off its hinges. The Viking was severely damaged and the Captain experienced difficulty with the elevator and

Figure 10.3 Bomb damage to BEA Viking, April 1950 (*Courtesy of Captain D. Arundel*).

rudder controls. After returning to and safely landing the aircraft at Northolt airport, Captain Ian Harvey was shocked to see holes on both sides of the cabin, large enough to permit unobstructed sight lines from side to side. The hole on the left side measured 5 feet by 5 feet 7 inches and the hole on the right side was even larger, measuring 5 feet 2 inches by 8 feet 2 inches. Other damage included the rudder control rod and the elevator and rudder trim controls which were severed, and the main elevator control which was jammed. The only casualty was a stewardess who was seriously injured and the criminals responsible for the sabotage were not traced. A motive for the crime was never discovered, and Captain Harvey was awarded the George Medal for his feat of safely completing the flight.

In 1972, a Yugoslav airliner en route from Copenhagen to Belgrade was ripped apart by a terrorist bomb while flying at 33 000 feet. All passengers and crew were killed except stewardess Vesna Vulovic, who, surrounded by the bodies of the other occupants and part of the wreckage, fell into deep snow in the mountains below and by some miracle survived. Her only injuries were to her right leg, and in 1986 she was still working for the airline JAT, but in a ground job.

On 23 June 1985, a Boeing 747 of Air India was flying normally at 31 000 feet, when suddenly, and without any emergency distress call being received, it disappeared from the radar screens in Ireland that provide air traffic services to North Atlantic air traffic. Underwater television pictures of the wreckage and its pattern of distribution, and those parts of the aircraft that were recovered, showed that the forward and aft cargo compartments ruptured before water impact. The fuselage section aft of the wings separated from the forward portion in flight and according to the Indian Commission of Inquiry, a bomb in the forward cargo hold had caused the disaster. Evidence that the flight had been normal for about five hours and that there was nothing on the Cockpit Voice Recorder to indicate an emergency, was regarded as support for concluding that sabotage was the probable cause of the disaster.

On 2 April 1986, there was an explosion on board a TWA Boeing 727 flying over Greece while on a flight from Rome to Cairo via Athens and at 11 000 feet descending into Athens. Four passengers were blown out of a large hole in the fuselage and were killed. The pilot of the B727 successfully landed the aircraft at Athens only 13 minutes later and the 118 passengers and crew members still on board the aircraft received no further injuries.

It was established that a plastic bomb had been placed in the passenger cabin and not, as might have been expected, in the cargo hold. Suspicion quickly turned to an Arab woman who had occupied seat F10 where the explosion occurred, on an earlier flight from Cairo to Athens on the same day, and it is suspected that she carried the bomb onto the B727 at Cairo by taking advantage of confusion arising from her late check in. The plastic bomb was probably the size of a pack of cigarettes and it may have had a plastic detonator which would have made it difficult to detect. It is probable that the bomb was hidden under the seat, perhaps in the container for the emergency life belt.

On 16 April 1986, US military airplanes attacked Tripoli and Benghazi, Libya as reprisals for acts of terrorism! On the following day, 17 April, El Al security at London Heathrow Airport prevented a bomb being smuggled onto an airliner (see earlier paragraph on El Al security).

In November 1986, a Thai Airways A300 aircraft flying from Manila to Osaka suffered sudden decompression when an explosion occurred in a rear toilet. The aircraft was successfully landed at Osaka, and an investigation revealed that a small bomb, probably a hand grenade, had exploded in a waste bin in one of the rear toilets, shattering the rear bulkhead and blowing out a venting hatch behind it. More than 40 small holes were discovered in the rear bulkhead and more holes up to half an inch in diameter were found in the ceilings and walls of the two toilet compartments.

At a later stage of the investigation, the safety ring and part of the trigger lever of an American M26 type grenade were found. Metal fragments of the grenade were removed from the body of Seiki Naka-gawa, a passenger, and from a stewardess who was walking towards the tail of the aircraft at the time of the explosion. Nakagawa, who was in the toilet at the time of the explosion, fell into the cargo hold and was considered lucky to survive. Other passengers reported that he had been drinking heavily, and the police, suspecting that he had carried the grenade onto the aircraft and had exploded it in the toilet compartment, arrested him. He is reported to have bought $900 000 worth of insurance before the flight.

In late November 1987, a Korean Airlines B707 disappeared while flying between Abu Dhabi and Bangkok, and all 159 persons on board lost their lives. Suspicion quickly fell upon a man and woman travelling on Japanese passports who had left the flight at Abu Dhabi, flown to Bahrein and changed their onward travel arrangements. When arrested for questioning at Bahrein, they took cyanide pills and the man died but the life of the woman was saved. It was proven that their passports were false, their journey had started in Belgrade and they had avoided security checks at Baghdad by remaining in the transit lounge before boarding the Korean B707. The woman was extradited to South Korea for questioning and trial for murder. In August 1988 President Zia of Pakistan and 30 other persons lost their lives in an airplane crash almost certainly caused by a gas bomb.

On 21 December 1988, a Pan American B747 was destroyed over Lockerbie, Scotland, killing 259 passengers and crew and 11 persons on the ground. A one kilogram Semtex bomb in the forward baggage hold separated the forward fuselage section of the airplane scattering wreckage over a wide area. It was established that the bomb was in a Toshiba radio/cassette player, contained in a suitcase and placed aboard the airplane at Frankfurt as joining baggage or as transfer baggage from a connecting flight from Malta, or loaded at London. Resulting recri-minations about this failure of security, blighted security relations between Germany, UK and USA and destroyed Pan American's repu-tation leading to its ultimate bankruptcy. In 1992, warrants were issued for the arrest of two Libyans employed by Libyan Arab Airlines at

Malta, and it was alleged that they used stolen Air Malta tickets to put a suitcase containing a radio/cassette player bomb aboard a flight from Malta to Frankfurt where it was transferred to the B747.

More than any other bomb outrage, this tragedy concentrated minds and led to improved aviation security in the USA, and internationally, with new measures being agreed at ICAO. It led to aviation sanctions being operated against Libya and to continuing speculation about other countries being involved – Syria and Iran. The publicity this disaster generated led to a huge increase in hoax bomb warning telephone calls in the UK, with more than 400 being made in one year!

In September 1989, a UTA DC10 on a flight from Brazzaville to Paris via N'djamena Chad, was destroyed over the Sahara desert killing 156 passengers and 15 crew, and the pattern of wreckage made it clear that an in-flight explosion blew the airplane apart. This tragedy attracted less attention than the Lockerbie disaster, but the French government reacted in a similar manner to the USA in the earlier event.

From the large number of incidents and different methods employed in the bombing cases on record, it is clear that the authorities responsible for security in civil aviation face a formidable task in seeking effective measures to end the threat of sabotage by bombings. On flights where a known threat exists, it may be necessary to inspect every item of baggage, cargo and supplies on board the aircraft, and the increased costs that will arise and the delays that are incurred, must be accepted as the price to be paid for safety (see Fig. 10.4).

Responses to bomb threats

In 1991, ICAO adopted a new Convention, 'The Marking of Plastic Explosives for the Purposes of Detection' requiring all manufacturers of plastic explosives to include a unique trace element in each batch of explosive to enable the source and date of manufacture to be established without doubt. It is believed that this information will assist in reducing illegal traffic in these materials and assist in the detection and arrest of traffickers and their accomplices. Documentation prepared for the ICAO Assembly in 1995, showed only 10 ratifications, approvals or accessions of 35 required before the Convention comes into force.

The FAA and specialist committees of ICAO, are researching the possibility of making airplanes 'bombproof' perhaps by lining baggage and cargo containers with blankets of high strength composites strong enough to protect the aircraft from the debris of the blast but porous enough to allow gases and increased pressure from the explosion to

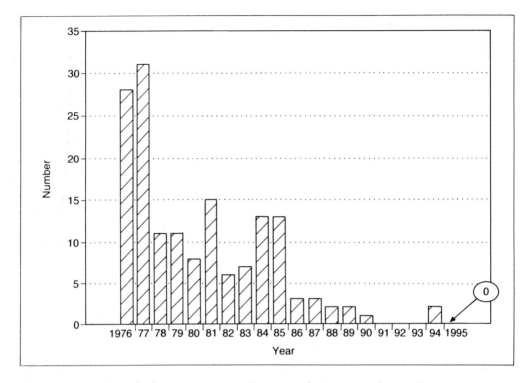

Figure 10.4 Incidents of sabotage 1976–1995 (*Courtesy of ICAO Annual Report*).

escape without damaging the aircraft. The containers would also release fire extinguishant. These hardened baggage containers would probably be used in conjunction with bomb detectors that would keep large bombs off airplanes, with the containers withstanding the effects of small bombs. A different and perhaps less practical proposal is a system that deploys metal cutters to open a precise hole in the fuselage to vent explosions overboard.

The security screening of airmail, baggage and cargo for bombs is a formidable task, and some reliance is placed on the 'known shipper' principle that releases carriers from screening anything if they have dealt with the shipper for a long period, but technological solutions necessary to screen bulk cargo is seen as a longer term desirable objective.

Responses to acts of terrorism

International responses to hijackings and other acts of terrorism are led by the UN and ICAO, with the former organisation concentrating on activities aimed at persuading member states to discourage these crimes by all political, legal and other means and to ensure the extradition or punishment of offenders. ICAO performs the detailed work required to develop the international Conventions that apply, and through its Committee on Unlawful Interference examines all other legal possibilities. On a technical level the ICAO Aviation Security Panel helps member states develop security systems aimed at preventing attempts at hijacking and sabotage from being successful.

Governments see hijacking and other acts of terrorism as an extremely serious symptom of a possible breakdown of international law, and are particularly concerned by the success of non-governmental organisations in gaining publicity for political objectives. These considerations and national policies result in some governments becoming leaders in the international fight against terrorism, particularly 'western' states that are most often the targets. The serious economic consequences of acts of terrorism against civil aviation is a significant unifying factor, and eventually the international Conventions aimed against the threat achieved more signatures and ratifications than any other Convention.

Terrorist attacks reflect international political trends and events, and the threat is subject to constant review and re-assessment by security forces who cooperate together for the common good. Interpol, the international police organisation, develops cooperation and intelligence sharing between national security forces.

Governments that maintain anti-terrorist security forces, conduct exercises to establish whether the forces are sufficiently well trained and equipped to meet the eventuality for which they were established. Some exercises are conducted in secrecy using only government personnel. On other occasions, exercises are conducted at major international airports and passengers are surprised and alarmed to be confronted by armed personnel and armoured vehicles.

Airlines vulnerable to terrorist action provide training for flight and cabin crews. Except for a minority that have a policy of resisting hijackers while in flight, the trend in training has been to teach crews to accede to the demands of hijackers, and to do so in a way that reduces the tension inevitably present in such circumstances.

Crew members are taught to recognise and discriminate between different psychological and behavioural patterns, and are briefed on earlier hijackings. For these purposes information gained by de-briefing hijacked crews is of great value. Crews are instructed that the only place to combat hijackers is on the ground using highly trained security forces, and are actively discouraged from taking 'heroic' action during the in-flight phase of a hijacking. They are directed to do nothing that may increase risk to passengers.

Pilot bans

At its annual conference in 1969, member associations of IFALPA agreed to seek to persuade states to sign, ratify and adhere to international conventions and/or make bilateral or multi-lateral agreements with like-minded states. Finally, if these courses of action failed to control hijacking, it was agreed to impose a 24-hour worldwide cessation of operations to highlight official inaction. This was to stress that hijacking is not a technical and safety problem to be solved within the civil air transport industry, but one for international law and united action by all governments through the UN and ICAO. The term 'Cessation of Operations', was chosen by IFALPA to forestall any attempts by governments and airlines to block pilot action by describing it as a strike, and thereby using anti-strike laws against the pilots.

The 'cessation' was to take place in late September 1969, starting at noon GMT and to last for 24 hours. Although it was clear that not all member associations would be able to participate, the Federation's leaders were agreed that the action would go ahead, and that it had a good chance of achieving the desired results. On 6 September, the leaders of IFALPA met the Secretary General of the UN, Mr U. Thant and were advised by him that it was essential for their purposes that a member state of the UN propose to the Security Council that the problem of hijacking be debated at the UN. The UN Representative of Finland was willing to take that action, but the USSR intended to use its veto to block the initiative. UN Representatives of Canada and the Netherlands then succeeded in having the item inscribed on the agenda of the General Assembly of the UN where a veto is not possible. The Federation's leaders called off the 'cessation' and were pleased by that early demonstration of the power of the international pilot group acting together on a matter where it had public support.

In June 1970, IFALPA was pleased when an Extraordinary Assembly of ICAO issued a declaration condemning hijacking and sabotage and approved recommendations establishing obligations and procedures in the event of an aircraft being hijacked to any country. ICAO recommended bilateral and multi-lateral extradition agreements between states, and offered its own 'good offices' as mediator in any problem that might arise between states in a hijacking situation. Throughout 1971 and 1972, IFALPA attached great importance to the initiatives of the UN and ICAO, and participated fully in the development of the Hague and Montreal Conventions. However, it was concerned that some states had not accepted their obligations under the Conventions. Greece had allowed an exchange of prisoners in return for the crew and passengers of a hijacked aircraft, the United Arab Republic and Jordan had received and welcomed hijackers and convicted saboteurs, and Israel had detained two Algerian passengers while they were in transit.

On 19 June 1972, the Federation carried out a 24 hour 'cessation' of operations. Although a number of member associations were not able to participate (in the case of USA ALPA the Supreme Court had issued a ruling on 18 June preventing its participation) it was nevertheless successful, and achieved its aim of generating activity at UN and ICAO levels, and also nationally to obtain ratifications of the international Conventions. At the end of 1972, IFALPA convened a Special Conference with the task of preparation for further meetings at ICAO, particularly of the Legal Committee. The conference decided to maintain pressure for international progress in the fight against terrorism, and to mount another 'cessation' of operations if dissatisfied with the rate of progress.

Progress was judged to be unacceptably slow, so another 'cessation' of operations for a period of 48 hours was arranged to begin from 1200 hours GMT on 25 October 1977. On that occasion greater support from the member associations was promised than ever before, perhaps because the hijackers of a Lufthansa Boeing 737, had shot and killed the captain on 17 October at Aden, before making the copilot fly the aircraft on to Mogadishu in Somalia. The stated objective of the Federation was to persuade the UN to call an urgent session to consider the adoption of an 'Enforcement Convention'. This would be used to apply pressure against any state that would not honour its obligations to the international community of nations with regard to crimes of unlawful interference with civil aviation.

By 19 October Mr Waldheim, the Secretary General of the UN, had agreed to call an extra session of the General Assembly, meeting as a Special Political Committee, to discuss the pilots' demands, and the Federation called off the planned 'cessation'. It was to be 1983 before IFALPA was again to flex its muscles in the cause of air safety, on the occasion of the destruction of Korean Air Lines Flight 007 by military forces of the Soviet Union.

At various times IFALPA has considered applying a ban against any state that it deems to be an 'Offending State' because of the state's action or inaction in cases of acts of unlawful interference with civil aviation. Among the criteria considered in applying that policy are the following:

- Failure to comply with the three international conventions
- Committing, or assisting in, an attack on civil aviation
- Failure to condemn or take positive action to prevent such attacks

■ Failure to take precautions against such attacks and to maintain a satisfactory level of security
■ Failure to release hijacked passengers and crew
■ Closure of runways or a failure to provide assistance to hijacked aircraft
■ Failure to treat hijacked aircraft as being in 'a state of emergency'

Among sanctions the Federation contemplates using in such circumstances are a selective ban against that state and/or the use of international publicity against the state. Both types of action have been imposed or threatened and are taken seriously by governments and international agencies.

The ending of the 'cold war' and an improved world political situation resulting in fewer hijacks of 'western' airliners, plus changing relationships between airlines and pilot associations – including a reduced number of nationalised airlines – make it unlikely that IFALPA will ever again call for a 'cessation' of operations. However, it can be satisfied that it played a major role in ensuring that crimes of violence against civil aviation received attention at the highest level.

11: Military Interception of Civil Aircraft

International law provides for a state to have sovereignty in airspace over its territory, and a right of self-defence including the interception of any unidentified aircraft approaching that airspace in order to establish its identity and intentions. In the event of a state establishing that the intent of the approaching aircraft is hostile, it has the right to take any action necessary to defend its territory and peoples. In recognition of problems that might arise from exercise of these rights, ICAO agreed in 1966 and again in 1973 to impress on all states the desirability of using interception procedures only as a last resort.

Interceptions of civil aircraft may occur when they approach a state's territorial airspace and are unidentified as a result of failures of the civil air traffic system. These failures could be caused by a breakdown of communications, a breakdown of coordination between air traffic control centres or an unintended navigational deviation by the aircraft.

The Chicago Convention developed two definitions for aircraft identity, state and civil, and stated in Article 13 that aircraft *used* in military, customs and police services shall be deemed to be state aircraft. All other aircraft are deemed to be civil and to be subject to the Chicago Convention and other regulations developed by ICAO and accepted by states. State aircraft of a contracting state are not permitted to fly over the territory of another state without authorisation, and states are to have due regard for the safety of civil aircraft when issuing regulations for state aircraft. It seems that the qualification *used* determines the state or civil status of an aircraft and presumably the status of the crew and passengers.

In some states it is common practice for civil airliners to augment military transport fleets at times of national need, and the legal position of civil crew members involved may be in doubt. On other occasions, persons in custody and deportees are carried in airplanes chartered for that purpose, or such persons are carried on normal scheduled flights, sometimes escorted by police officers and perhaps making the aircraft a 'state' aircraft and thereby affecting the legal status of the flight, other passengers and crew members. IFALPA believes that where doubt exists, the status – civil or state – should be determined by a competent authority before the flight takes place.

A civil aircraft carrying military personnel or police could be deemed to be a 'state' aircraft and that interpretation became important in October 1985:

- An Egyptair airliner was being used to transport the alleged hijackers of the Italian cruise ship Achille Lauro, and the flight crew believing they were operating a civil aircraft on a VIP charter to the Egyptian Government, filed a civil flight plan as required by ICAO rules. While en route and flying in international airspace, the aircraft was intercepted by armed aircraft of the US Navy and forced to land at an aerodrome in Sicily where the alleged hijackers were removed and detained, after which the airliner was permitted to depart. No action was taken by ICAO in respect of this interception in international airspace, and it is believed that diplomatic contacts resulted in an informal decision that the airliner was being used in police service at the time of the occurrence, and was therefore a 'state' aircraft.
- In September 1984, a USA registered B707 operated by South Pacific Airways over a Polar route, was carrying 120 Fijian military personnel to the Middle East to serve in an international peace keeping force, when it became lost and penetrated Norwegian airspace without authorisation. The subsequent investigation was conducted under civil procedures although the flight was performed solely for military purposes.
- In February 1986, a Grumman G2 aircraft operated by Libyan Arab Airlines departed Tripoli for Damascus carrying Syrian diplomats returning from a conference. The flight was intercepted by Israeli military aircraft while flying in international airspace and forced to land in Israel for the purpose of checking the identities of the passengers. No irregularity was found, the aircraft and passengers were released and ICAO subsequently censured the State of Israel for its actions.

This inconsistency of treatment between civil and state, was brought to the attention of ICAO by IFALPA, which seeks clarification of the status of aircraft, crews and passengers when the *use* of an aircraft could be argued to be that of a state aircraft. The Federation wants to know whether civil or military procedures would be followed in accident investigation, whether special permission is required to overfly the territory of another state and whether such an aircraft could be intercepted in international airspace and be made to land. IFALPA points out that passengers aboard an airliner may be a mix of civil and military personnel, and civil passengers could find themselves in an unpleasant situation if some unfriendly state decided to treat the aircraft status as 'state'. Questions of insurance liability arise and the limits of liability of the carrier in the event of an accident might be affected. For all these reasons it seems desirable for the legal questions raised by IFALPA to be clarified, for the safety of civil aircraft, passengers and crew members may be adversely affected by the present uncertainty.

Interception procedures

An internationally agreed interception procedure devised by ICAO, and subsequently modified in the early 1980s, takes account of changing circumstances such as the high performance of civil aircraft and experience of actual interceptions. Although the procedure is imperfect because of the difficulty of conducting communications between civil and military aircraft of different nationalities, it is believed that its universal use should prevent any disaster caused by military action against unarmed civil aircraft. However, actual interceptions resulting in a high level of risk or disaster include the following:

■ In 1952, an Air France aircraft was intercepted by Soviet fighters while flying outside the approved air corridor and over an East German chemical manufacturing complex near Berlin. Soviet interceptors used internationally agreed procedures, but the Air France pilot attempted to evade the interceptors by flying into cloud and toward the air corridor. A cannon shell hit the aircraft, missing the fuel tanks by only six inches but the aircraft was safely landed at Berlin.
■ The Israeli Air Force shot down a Libyan Air Lines B727 over Sinai in 1973.
■ In April 1978, a Korean Air Lines Boeing 707 strayed into Soviet airspace near Murmansk, hundreds of miles away from its planned route from Western Europe to Anchorage, Alaska. It was fired upon by a Soviet SU15 military aircraft and made a forced landing.
■ In September 1983, a Korean Air Lines Boeing 747 was shot down while on a flight from Anchorage, Alaska, to Seoul, South Korea, resulting in the deaths of all of the 269 people on board. Again the Korean airliner was hundreds of miles away from its scheduled route.

■ In July 1988, Iran Air Flight 655, an Airbus A300 on a scheduled flight over the Arabian Gulf was shot down by missiles from a US Navy surface vessel with the loss of 290 lives.
■ In February 1996, Cuban military airplanes shot down two US-registered civil light aircraft, and the ICAO Council adopted a resolution deploring the incidents.

Each of these events caused major political and technical attention to be given to the basic problem of how to prevent such disasters. ICAO reviewed interception procedures, and in 1980 after the first 1978 Korean Air Lines incident, sent proposed modified procedures to member states and concerned international organisations, for comment. The reply made by IFALPA was that it agreed the modified procedures were an improvement, but it had 'serious misgivings' with regard to the following main points:

■ Guidance material status of the modified procedures should be up-graded to Regulatory status.
■ Material concerning the use of weapons should include specific reference to the use of guided missiles.
■ The civil emergency radio frequency of 121.5 mHz and Secondary Surveillance Radar (SSR) should be provided to all intercepter aircraft and their ground control units to facilitate communication with intercepted civil aircraft.
■ Navigation capability standards set by ICAO should be reviewed and improved.

Had these criticisms and suggestions been accepted by ICAO and its member states, the disaster that was to befall the Korean Air Lines B747 only three years later might have been prevented.

A weakness of the ICAO procedures is use of substitute methods of communication to compensate for the absence of direct radio communication between military pilots/controllers and civil pilots. These substitutes include unsatisfactory items such as 'rocking wings', flashing navigation lights at irregular intervals, lowering the landing gear, etc. They also require the intercepting aircraft to fly in close formation with the intercepted aircraft, particularly in early stages of the interception so as to indicate that an interception is being made.

ICAO procedures attempt to provide for occasions when radio communication is possible, but a common language is not available, with instructions being passed to the intercepted pilot by the use of standardised phrases designed by experts in linguistics so as to be understandable to most nationalities and useable in most circumstances. An example being a Russian fighter pilot intercepting an airliner flown by a Korean.

The most important ICAO rule affecting interception procedures is that contained in paragraph 7.1 of Annexe 2, Rules of the Air:

'Intercepting aircraft should refrain from the use of weapons in all cases of interception of civil aircraft.'

The policies of IFALPA similarly require:

'When ... a civil aircraft inadvertently strays from its correct flight path and infringement of prohibited airspace appears likely or actually occurs, the situation should be dealt with on the basis that the civil aircraft requires assistance, as distinct from police action. Invariably, the aircraft concerned will be in some form of distress, ranging from being lost to being in an emergency condition.'

Had these complementary items achieved worldwide acceptance, disasters experienced over a period of more than 40 years need not have occurred. A surprising factor, uncovered in the aftermath of the ICAO investigation into the KAL 007 disaster of 1983, was the appalling record of states with regard to implemention of the ICAO recommendations on interception procedures. Of the 151 member states:

- Only 61.5% published information relating to the procedures in their own Aeronautical Information Publications (AIPs)
- 37.7% published differences from the relevant ICAO provisions
- 23.8% had no differences from the relevant ICAO provisions
- 38.5% had published no information in their AIPs concerning compliance with *or* differences from the relevant ICAO provisions
- 46.3% had not published any procedures or visual signals relating to interception as required by the Standard in Annexe 15 of the Chicago Convention to which all member states are party.

It was abundantly clear that many states had taken little interest in the matter and had not ensured that their airlines and pilots had the information required to enable them to operate safely in 'sensitive' areas of the world where interception may occur. ICAO defined 'sensitive areas' to include all areas where major political tensions exist, and elimination of risk of interceptions must therefore await the solution of difficult and long lasting political problems. It seems certain that safety problems caused by military interception of civil aircraft will exist for many years to come.

Korean Air Lines Flight 007

On 1 September 1983, a shocked world learned that a Boeing 747 with 269 persons on board and operated by Korean Air Lines was missing on a flight from Anchorage, Alaska, to Seoul, South Korea. Early confusion arose from a marked reluctance of Soviet authorities to release information about the disaster, until the rest of the world had become convinced from reports made by the media and statements made by the Governments of the USA, Japan and South Korea that the aircraft had been shot down by military aircraft of the Soviet Union. The disaster took on greater significance than it might otherwise have done because of the political relationship between the Soviet Union and the other directly concerned governments.

The USA published 'transcripts' of communications between the Soviet military controllers and the intercepting fighter pilots, and these were widely accepted as confirming that the shooting down of KAL 007 was a deliberate act. At a later stage of a subsequent investigation, it became clear that the 'transcripts' were not an accurate record because of the imprecise translations made of the conversations, but most of the world remained convinced that the USSR had committed a major crime.

Calls were made for action to be taken against the USSR, but the UN and most governments are not equipped to make quick and effective responses in such circumstances, and so the first practical response was made by IFALPA. By 9 September the Federation had consulted member associations operating flights to the USSR, and obtained the agreement of almost all of them to apply a ban on flights to Moscow from 12 September. In an open letter to the world's press, the President of IFALPA explained that the Federation's action was motivated by the 'traditional and legal responsibility of airline pilots for the safety of their aircraft with their passengers and crew'.

The action taken by IFALPA was widely popular, and gained practical support from other unions including the air traffic controllers of Norway, who refused to issue air traffic clearances to flights to and from Moscow. Although announced as being for a period of 60 days, IFALPA recognised that it would be difficult to maintain a ban that directly affected only 16 of its members. Of the sixteen, pilots' associations of Austria, India, and Yugoslavia were prevented from supporting the ban by their governments, and the pilots' association of the USSR expressed its disagreement with the ban. Some West European Governments later announced their own form of ban on flights to the USSR, but the action of the world's airline pilots was the first, and it had a major effect on the perceptions and actions of a number of national and international organisations.

At a special meeting of the ICAO Council, most states' representatives condemned the action of the Soviet Union and demanded that the organisation conduct its own investigation into the incident. The representative of the USSR expressed his state's sympathy for the victims of the disaster, but alleged that the flight of KAL 007 was 'a sanctioned and organised provocative flight'. The Council's final statement deplored the destruction of the aircraft, urged the USSR to assist the bereaved families to visit the site of the accident and return the bodies of the victims and their belongings promptly. It also urged an immediate and full investigation of the 'said action' and an improvement of interception procedures.

Taking note of these decisions, IFALPA lifted its ban on 3 October, pleased with the results achieved for it had intended only to make an effective demonstration of its concern over the shooting down of a civil airliner. It knew that a ban on flights to Moscow could be circumvented by taking a flight via another East European city with onward travel by the USSR airline Aeroflot, and could be no more than a minor inconvenience to the USSR, but it confidently and correctly expected this action to have a great effect on public opinion.

The ICAO report into the circumstances of the shooting down of KAL 007 was thorough, and made a number of recommendations aimed at preventing similar incidents. The USSR continued to assert that the airplane may have been on a spying mission, but most governments accepted the view of the USA, that the shooting down was a criminal act. IFALPA stated that criticisms made of the Korean flight crew in the report were based on assumptions and not evidence.

An Extraordinary Session of the ICAO Assembly from 24 April to 10 May 1984, considered possible amendments to the Chicago Convention aimed at making disasters such as the KAL 007 less likely. Emphasis was placed on having states refrain from the use of force against civil aircraft and suitable Resolutions were drawn up and approved. The proposal of IFALPA for direct radio communications between the intercepted pilot and the intercepter, or his controller, was not agreed.

Eleven years after the ICAO Special Assembly agreed 'to reaffirm the principle that every state must refrain from resorting to the use of weapons against civil aircraft in flight' only 80 of the 102 signatures required to bring the amendment into force, had been achieved!

IFALPA proposed improved coordination of activities potentially hazardous to civil aircraft (war zones etc.) be introduced. However, more needs to be done to ensure the safety of civil airplanes at times of political and military confrontation, as shown by reports in August 1995, that two wet-leased Russian IL – 76 aircraft flying to Kabul, Afghanistan were intercepted and made to land at Kandahar by two Mig 19s operated by a dissident group. Subsequently an Afghan opposition alliance warned all foreign airlines not to fly over territories under its control, and IFALPA advised all operators to avoid flying through Afghanistan airspace and to follow contingency plans drawn up by ICAO. It is known that many airlines continued to fly in the affected airspace.

It was not until 1992, after changes of leadership in Russia, that President Yeltsin handed the KAL 007's flight data and cockpit voice recorders to the South Korean government apologising for the tragedy, which he said was due to the 'criminal regime' of the former USSR. As recently as May 1996, dependents of passengers killed in the tragedy were seeking compensation, and a US Federal court awarded $10 million to the family of a passenger, stating that the normal limit of $75 000 did not apply because of the crews' 'wilful misconduct'.

Iran Air Flight 655 (July 1988)

US military authorities and ICAO investigated this tragedy with the USA accepting liability and paying damages to the dependents of those who lost their lives. The incident happened at a time of tension in the Arabian Gulf with an appalling lack of coordination between military and civil authorities. Civil pilots on scheduled flights were being challenged by military authorities, e.g. 'unidentified aircraft at 20 000 feet, range 15 nautical miles, bearing 310 degrees, this is US warship, state your nationality and intentions'. This message is ambiguous as the pilot of the aircraft being challenged does not know where the ship is, and may not realise the message is for him or her. Procedures were improved after consultation between IFALPA and the US authorities.

12: Advanced Technology

When examining the effect of technological developments on air safety, it is worthwhile to consider the most advanced airliner in production (as of 1996) to see how designers, engineers and test pilots have addressed the man-machine interface and safety problems that existed with earlier aircraft. The effects of advanced technology on flight operations and the operating environment are also considered.

Boeing 777

This large twin-engined airliner made a first flight on 12 June 1994, was certificated in 1995 and the first revenue-earning flight was made by United Air Lines from London to Washington DC on 7 June 1995. Models are available with either Pratt and Whitney, General Electric or Rolls Royce engines, with thrust levels ranging from 77 000 to more than 90 000 pounds with the more powerful engines to be used for later versions of the aircraft. For the first time, an aircraft manufacturer involved major customer airlines and suppliers in the design process, and the final product owes as much to what the airlines did not like about earlier aircraft they had operated, as to perceived desirable qualities of the new. Engineers, pilots, cabin crew, sales personnel, and cargo handlers worked together in a range of design committees, and about 300 customer airline staff temporarily participated in 'design/build' teams at Seattle. All are likely to be pleased with the results!

Boeing states that the 777 is its first aircraft designed from the inside out, with the passenger cabin forming the start point for the whole structure, and having a cylindrical and not ovoid or 'double bubble' fuselage. The aircraft sits neatly in size between the B767 and B747, and is able to carry up to 440 passengers in a high density all-economy seating version. Range varies from 4000 to 6000 nautical miles according to payload, and later models are likely to rival the B747–400 with ranges greater than 7000 nautical miles.

Structure

Carbon Fibre Reinforced Plastic (CFRP) composite materials make up about 10% of the structural weight of the 777, and offer weight reduction, resistance to fatigue and corrosion, resistance to impact damage and reduction in need for structural repairs. They are used for vertical and horizontal stabiliser main torque boxes, fuselage floor beams, engine cowlings, flight control surfaces, landing gear doors and wing to body fairings. Materials used are resistant to standard aircraft liquids such as hydraulic fluid, jet fuel and deicing fluid, with service proven finishes applied for protection against moisture, ultraviolet light and erosion damage. In preparation for this increased use of composites, Boeing built five B737 aircraft with composite horizontal stabilisers which have been in airline service since 1984 with good results. Of particular importance for safety (and economics) is the fatigue insensitivity of composite structures, and Boeing has developed an initial minimum structural inspection program certified by the FAA.

B777 certification program

Full-scale static and fatigue structural tests took place to meet the requirements of the FAA and JAA regulatory authorities, and the wing was tested to destruction. For the static tests, more than 4300 strain gauges were fitted, and more than 500 miles of cables connected these gauges to a data acquisition system. The wing destruct test took place on 14 January 1995, with both wings failing almost simultaneously at 103% of proof load at about 50% span length, with the largest measured wing-tip deflection being more than 24 feet. Fatigue testing at a rate of every 5 minutes for 24 hours per day continued until late 1996, and by early July 1995 more than 25 000 cycles were completed.

The test program meets all requirements of the FAA and JAA's and involves approximately 4800 regulations. A comparison between key test parameters of the certification programs for B767 and B777 airplanes is shown in Table 12.1.

Extended-range twin-engined operations (ETOPS)

Urged by customer airlines, Boeing sought to have the B777 certificated for 180 minute ETOPS operations from day 1 of receiving a type certificate, and this required an extensive test program beyond those required

Table 12.1 B767 and B777 certification programs.

	767	777
Test airplanes	5	4
Program duration (months)	10	10.5
Number of flights	1584	1461
Flight-test hours	1793	2092
Ground test hours	667	3236

for gaining a type certificate. Previous practice of regulatory authorities had been to require x years and y hours of trouble-free airplane/engine combination operations, before giving approval for ETOPS flights. However, Boeing set out to meet the requirements by means of basic system design features and a unique test program. The baseline B777 systems meet ETOPS requirements without modification, with a new feature being an air turbine starter in addition to an electric start system for the auxiliary power unit (APU).

Boeing arranged that a single airplane fly a minimum of 1000 flight cycles to demonstrate the necessary reliability. (A flight cycle includes an engine start and taxi, a take-off, climb, cruise, descent, approach, landing and engine shutdown). During this 1000 cycle validation program, the airplane was required to be maintained and operated like a 'mini-airline', including the use of in-service airline procedures and manuals. Starting in December 1994, the airplane accumulated about 1800 hours during the 1000 cycles, with the flight experience addressing more than 5000 individual requirements. Flight time varied from less than an hour to more than ten hours, including eight 180 minute single engine diversions. These eight single engine flights are said by Boeing to equal 24 total diversion hours accumulated during the first five years of B767 ETOPS operations.

The airplane was taken to various remote bases to encounter a mix of environmental conditions including exposure to temperatures ranging from minus 44°F to plus 108°F, and the first customer airline United, performed 90 of these flight cycles over its own routes including Washington–Los Angeles, Denver–Miami, and Washington–Honolulu. The FAA granted 180 minute ETOPS approval to United for the B777/PW4084 airplane on 30 May 1995. British Airways with the GE powered airplane, and Cathay Pacific with the Rolls Royce engine will achieve the same results after similar programs.

Opposition to this unique ETOPS certification program came from pilot associations who, while admiring the quality of the Boeing program, do not accept that it can duplicate or match the previously required 2 year and 'x' engine hour demonstration by airline-operated aircraft formerly required by the FAA and JAAs.

Airplane information management system (AIMS)

This computing system is described as an 'integrated modular avionics concept', and occupies two cabinets located in the main equipment center. There are four input/output modules (IOMs), and four core processor modules (CPMs) in each AIMS cabinet. AIMS is used to support the following systems:

- Primary display system: controls, formats and shows information on airplane data, navigation data, flight plan data, maintenance page data, engine indicating and crew alerting data, check lists, and communication data

- Central maintenance computing system: collects, stores and displays maintenance data for most of the 777 systems

- Airplane condition monitoring system: monitors, records and reports information for maintenance, performance, troubleshooting and trend monitoring data

- Flight data recorder system: records mandatory and optional flight data for the most recent 25 hours of operation

- Data communication management system: supplies Aircraft Communication and Reporting System (ACARS) datalink, flight deck printer support, and avionics local area network interface to AIMS

- Flight management computing system: for navigation, flight planning, performance management and navigation radio tuning

- Thrust management system: moves thrust levers when commanded, gives thrust limit displays and auto throttle modes during take-off, and all flight phases and supplies trim commands to the engines.

Airplane maintenance

A new system devised by Boeing, provides on-line access to airplane engineering drawings in a two-dimensioned digitised format that replaces an earlier system of providing up to 247 000 of 35mm aperture cards (for a B747–400) that eventually become unreadable due to frequent use.

With the new system, airline computer work stations and printers are connected to Boeing's 750 gigabytes data base. With 'zoom' and 'cut and paste' techniques, the end user has access to the latest drawings.

777 onboard maintenance system

The 777's onboard central maintenance computer (CMC) is the latest step in the development of computer based maintenance aids. It contributes to effective fault isolation of digital based components and systems, and allows maintenance management of redundant design features to ensure airplane availability. As Boeing's Chief 777 Mechanic states, 'you can't troubleshoot a computer chip by looking for physical evidence of failure ... broken chips don't leak, vibrate or make noise. Bad software within them doesn't leave puddles or stains as evidence of misbehaviour'.

Digital built-in test equipment (BITE) permits many parameters within a line-replaceable unit (LRU) to be monitored, and to generate fault reports whenever one of these parameters is out of acceptable limits. BITE monitors and BITE logic in different airplane systems report to the flight deck and the CMC, with engine indication and crew alerting system (EICAS) on the flight deck displaying an appropriate message alerting the pilots to an abnormal condition and the CMC setting maintenance messages for about 87 airplane systems. There are provisions for connecting a portable maintenance access terminal (MAT) at the nose landing gear, main gear, electronics compartment, flight deck and stabiliser compartment. Data can also be accessed from two optional flight deck displays (one by each nose-wheel steering tiller) with a MAT positioned at the right rear of the flight deck aisle stand. The mechanic will communicate with the CMC via the MAT screens, a cursor control device and, if required, a keyboard and will see a menu-structured display.

Flight controls

The B777 has fly-by-wire flight controls but its systems provide visual and tactile clues for pilots that are missing from Airbus aircraft, with trimming controls and the two sets of conventional control columns, rudder pedals and thrust levers moving in both manual and automatic flight modes. Control columns are connected to computers that signal actuators at the flight controls, but 'feel' and control forces are as good as the best of other aircraft. The primary flight control system is based on a triplex system of primary flight computers (PFCs) with data flows on three independent but identical channels – left, centre and right. A significant difference from Airbus practice is that the three identical software programs were written by a single team, whereas Airbus used three teams to develop a version of the software each, to reduce the chance of 'bugs' affecting all three. Only airline experience will tell which is the best practice!

Each computer inputs commands into only one of three data buses but receives data from all three. A PFC contains three independent dissimilar lanes; one in command, one monitoring and one on standby and it is expected that aircraft will be able to remain in service with one lane of a PFC unserviceable. Electrical power supplies from four engine driven generators, an auxiliary power unit (APU) that can be re-lit at maximum altitude after a prolonged cold soak, an emergency ram air turbine (RAT) and batteries, provide an ample level of redundancy.

The system is 'smart' in that it provides flight envelope protection, including stall protection, overspeed protection and bank angle protection. It provides stability augmentation and compensation in each control axis – plus asymmetric thrust compensation. Instead of giving the automated flight control system authority to intervene and override pilot inputs as in the latest Airbus aircraft, Boeing provides ample tactile, aural and visual warnings to a pilot that the aircraft is near the boundaries of a normal flight envelope, by means of aural warnings, flight instruments and greatly increased control forces. A pilot may accept the warnings or ignore them, and fly manually to extreme boundaries in order to use maximum performance in an encounter with windshear or to avoid collision with terrain or another aircraft etc. Out of trim control forces increase with speed as in orthodox control systems, and the Boeing aircraft is governed by speed stability not pitch stability as in the Airbus fly-by-wire systems. 'If you pull back on the stick and let go, the column will go back to where it was.'

There are three operating modes for the primary flight control system:

■ Normal   when the PFCs supply all commands, and full functionality is provided

■ Secondary   when functionality (and protections) is reduced. This mode is entered automatically when there are sufficient failures in the system so that normal mode is not supported

■ Direct   available by selection or entered automatically due to total failures of PFCs or data busses. It uses mechanical connections from the control wheels to operate Nos 4 and 11 spoilers and from the stand-by pitch trim handles to the stabiliser. It is intended that this mode is used only until electrical power is restored, as it removes all protections and trim bias for configuration changes

Thrust asymmetry compensation (TAC) automatically applies rudder when thrust asymmetry exceeds 10% of rated thrust, in both manual and automatic flight. TAC operates at all airspeeds above 80 knots, and on the ground during take off, except when using reverse thrust. Residual yawing moment and a displaced rudder pedal enables pilots to positively identify which engine may have failed. To ensure that TAC does not mask an engine failure on take off, the system is designed to counter most but not all of the airplane yawing moment. Once in the air TAC fully compensates for the engine-out condition. Three auto pilots are provided with all three being simultaneously engaged by either pilots' control switch.

Very good flying qualities with built-in safeguards, and excellent flight deck layout including generous-sized flight deck windows should be a significant factor in preventing accidents that have pilot handling as a causal factor. Representatives of pilot associations who have flown Boeing and Airbus fly by wire aircraft are of the opinion that pilots transitioning from orthodox flight control design airplanes will find fewer difficulties with the Boeing design.

Flight deck displays

Six large flat panel displays are used for two primary flight displays (PFDs), one engine indicating and crew alerting system (EICAS) and one multi function display (MFD). Two further MFDs are normally dedicated to navigation displays but all three MFDs can be selected to show any of the following displays: navigation, secondary engine instruments, electronic check lists, status displays, system synoptic displays or communications (integrated data link). A few airplane systems provide an automatic display of non-normal conditions, e.g. low engine oil pressure.

The PFDs and NDs are side by side before each pilot and the EICAS and MFD are one above the other in the center. After a long period of development, active matrix liquid crystal displays (AMLCDs) replaced

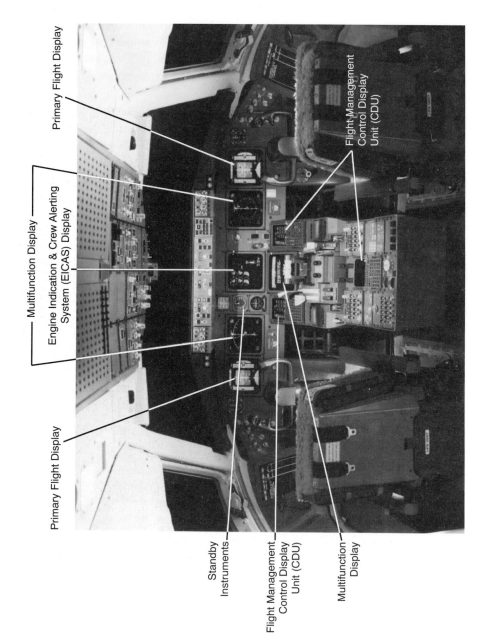

Figure 12.1 (a) B777 flight deck displays (*Courtesy of Boeing*).

Optional Side Display
(first officer's shown)

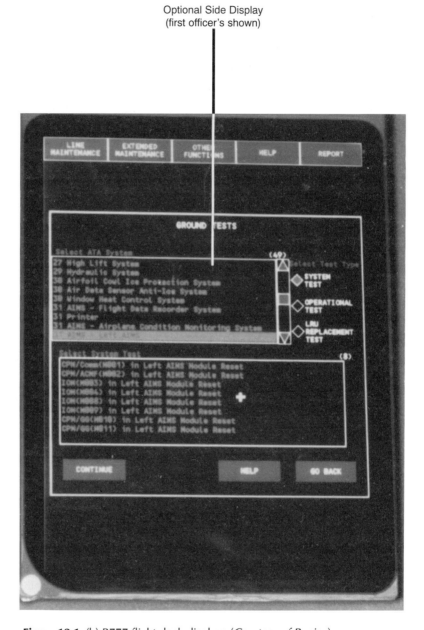

Figure 12.1 (b) B777 flight deck displays (*Courtesy of Boeing*).

Maintenance Access Terminal (MAT)
(right rear side)

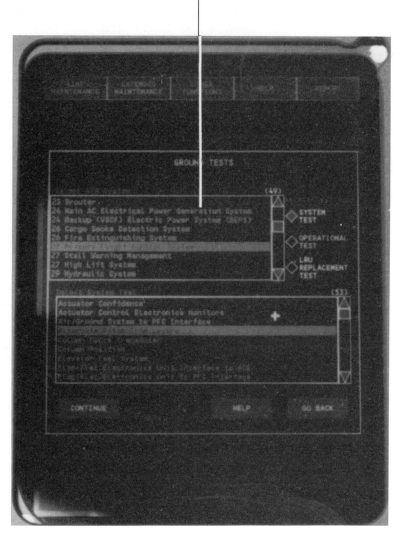

Figure 12.1 (c) B777 flight deck displays (*Courtesy of Boeing*).

cathode ray tubes fitted to the B747–400, and are said to consume less power, to weigh less, take up less space and have a higher reliability. Human factor experts used a B777 flight deck engineering simulator capable of simulating the entire ambient lighting environment encountered by airplanes, and established that pilots require as much as 2000 times more light from displays during daytime conditions than at night, and that controls must permit display intensity to be turned low enough so the pilot can see targets (other aircraft, runways etc.) out of the window.

In addition to the six ALCMDs, pilots have in their fields of vision, glare shield controls, stand-by flight instruments, left center and right flight management control display units (CDUs) and a flight deck printer. Optional displays are a maintenance access terminal and two side displays. The B777 flight deck displays are illustrated in Figures 12.1(a), (b) and (c).

Flight training

Flight training courses planned for pilots are as follows:

- Academic and part-task skills are taught using Computer Based Training (CBT)
- Integration of part-task skills, procedures training and basic Cockpit Resource Management (CRM) are taught in the Fixed Base Simulator
- Airplane handling, maneuvers training and CRM are taught in the full flight simulator
- Skills and confidence are verified and a check ride conducted in the 777 airplane

Courses vary according to whether pilots have 'glass cockpit experience' with those new to the systems taking additional CBT and simulator flights. After systems knowledge and procedural skills are acquired, students proceed to the full flight simulator where instructors are type rated, fully-qualified pilots. After that stage of the course is complete, students transition to the airplane.

Aerospace technology of the future

The design and construction of airline aircraft will be evolutionary and not revolutionary for the next 20 years, and any 'great leap forward' in performance will await allocation of financial resources by the military. Concorde may remain the only supersonic airliner for the whole of the period and subsonic aircraft development will concentrate on improved economic performance. Lighter aircraft structures and improved engine fuel consumptions will enable greater payloads to be carried at lower costs, or alternatively, the range of the newly developed aircraft to be increased with long range aircraft operating non-stop and profitably between city pairs that are more than 8000 miles apart. A two pilot crew will be standard, with extra flight deck crew being carried only when duty times exceed limits at which considerations of crew fatigue apply. It is probable that research and development programs will be aimed at: propulsion systems, energy forms, automatic control systems, noise suppression, computers, new materials and air traffic management as these are likely to produce the greatest economic returns for the aerospace industry.

Each new aircraft type will be safer than preceding types, and provided sufficient resources are allocated to regulatory safety bodies, there will be continuous improvement in airline safety per flight or mile as measured by statisticians, although the growth of air traffic may result in more accidents and deaths each year.

Subsonic aircraft

It is probable that subsonic airliners in the medium-term future will be similar in appearance and performance to those in service in 1996, typified by the Boeing 777 which merits a separate and special description because of its advanced technology. Aircraft will almost certainly cruise at speeds of less than Mach .87, because higher speeds have adverse effects on fuel consumption and costs, but there is no absolute limit to the size of aircraft that can be built other than difficulties in handling at airports and in utilising the volumetric capacity made available by a very large hull. Design efforts will be aimed at reducing capital and operating costs and noise nuisance, with promising developments in sight from new materials that will reduce structure weight and maintenance costs and extend safe operating life.

In late 1993, British Airways stated a requirement for a new large airplane to cope with expected passenger growth rates of 5–6% per year. 'By the turn of the century, the B747 will be too small on routes where frequency increases are not possible.' The requirement went on to state the following objectives:

- Six hundred seats
- Able to use all airfields used by B747 airplanes
- A base turn round time of 105 minutes
- No major maintenance for ten years
- Engine changes of 30 minutes
- Same night noise classification, wake turbulence separation and runway occupancy times as the 747
- Improved system reliability and redundancy plus 100% accurate fault diagnosis
- 20% reduction in direct operating costs compared with the 747 obtained by reduced fuel burn per passenger kilometer, longer service life, reduced down time and lower lifetime costs

A possible way to reduce fuel burn is to adopt a new drag reduction technique of 'riblets' which are microscopic grooves aligned with the airflow. In limited flight trials, adhesive plastic film with grooves 0.002 inches deep and wide, have shown a possibility of achieving drag reductions of 3–4% for airliners operating at Mach 0.8. Laminar flow engine nacelles are another possible means of reducing drag and fuel burn. By 1996 it seemed that an intermediate stage between the B747 and the New Large Aircraft (NLA) would be necessary and likely to be met by a 'stretch' of the B747.

Aerodynamic and structures research indicate that it is possible to reconfigure wing shapes while in flight to the optimum for current operating conditions, the so-called 'mission adaptive wing', so finally escaping the tyranny of an aircraft wing being an ideal shape and size for only one brief period in any particular flight. Variable sweep wings are in wide use for military aircraft required to perform military tasks at both sub and supersonic speeds, and were proposed for the USA's supersonic SST aircraft that was to be built by Boeing in the 1970s but was cancelled at an early stage of development. When eventually used in airliners, variable sweep could improve the safety of take off and landings by reducing the minimum safe airspeed thus shortening take off and landing distances.

Supersonic aircraft

Operating experience gained with the 1960s technology Concorde (see Figure 12.2), has demonstrated that supersonic flight is safe and popular with that segment of the passenger market for who time is extremely valuable. Concorde could not have operated profitably had not the huge

Figure 12.2 Concorde (*Courtesy of British Aerospace*).

development, testing and manufacturing costs been written off by the British and French governments, but it is highly unlikely that the financial measures adopted for Concorde will again find favour. The route network available to Concorde is restricted because of supersonic boom, and that problem will remain for future supersonic aircraft unless they fly much higher, or some other means is found to reduce sonic boom nuisance. Concorde's major achievement has been to prove that supersonic flight is safe, for in the 25 years since its first flight, it has not injured or killed a single passenger. This is a remarkable feat for an airplane that flies so fast that the wing is 1 degree down in a turn to the left to compensate for the rotation of the earth below, and where 1.5% of airplane mass is supported centrifugally. Low utilisation of Concorde airplanes will enable them to fly well into the 21st century, and the operational conditions of a dry stratosphere and aerodynamic heating of the hull ensure that there is virtually no corrosion.

Studies for a successor for Concorde have been undertaken in the USA and in Europe, and it seems to be agreed that a new SST must be able to carry at least three times Concorde's payload, have twice the range at fares comparable to today's subsonic airplanes and be environmentally friendly. In 1996, NASA released information about a preliminary design concept for an M 2.4 four-engined double delta winged airplane, with a maximum weight of 335 000 kg, a length of 100 m, a wingspan of 40 m, and three class accommodation for 310 passengers. The wing leading and trailing edges, unlike Concorde, have moving lift/control surfaces, the forward fuselage has width for six abreast seating and the main cabin for five abreast. Range is to be 9200 km and the noise target is Stage 3 minus 3 db. An Enhanced Visual System (EVS) is to be provided to remove the need to 'droop' the nose to improve vision for take off and landing.

In 1995, NASA and a group of US manufacturers arranged with Tupolev, the buiider of the Tu144 (see Figure 12.3) the former USSR's supersonic transport (SST), to use an example of that airplane as a flying laboratory. Funds for this project are wholly from US sources and it is planned to investigate the following: surface, structure and engine temperatures, sonic boom signature, boundary layer, the wing's ground effect, aircraft handling, noise (interior and exterior) and the way the structure flexes in flight. A total of 32 test flights are planned, all to be flown over Russia by Tu pilots, and the airplane selected for the program is 77114. This is one of the D models which will be fitted with engines originally developed for the Tu160 Blackjack bomber, and expected to increase speeds from the original Mach 2.15 to Mach 2.35. The airplane

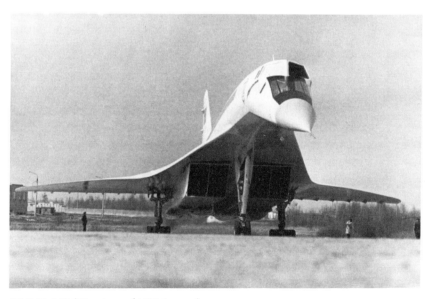

Figure 12.3 Tu144 (*Courtesy of MIR Agency*).

will be fitted with a new digital data system plus thermocouples, pressure sensors and skin friction gauges to measure the boundary layer. Costs are estimated to be only $14 million – a small fraction of the $440 million NASA contract awarded to Boeing and McDonnell Douglas for high speed research. During the bilateral discussions that lead to this historic agreement, it was learned that Tupolev still dreams of building a next generation SST – the Tu244.

Hypersonic aircraft

Should proposals being put forward in the USA come to fruition, boundaries between flight in the earth's atmosphere and space travel will be crossed. In 1984, President Reagan announced support for a vehicle that would give the USA a single-stage-to-orbit capability, providing repeated access to space for about one-tenth the cost of the Shuttle. It is proposed to take off and land horizontally from conventional runways using air breathing engines to achieve the hypersonic speeds necessary for orbital flight. The vehicle is named National Aero Space Plane (NASP) and a civil version would become the *Orient Express* capable of flying from Los Angeles to Tokyo in just over three hours at Mach 2.7.

The program's objectives may result in an airliner cruising at speeds between Mach 5 and Mach 8, a hypersonic military reconnaisance aircraft flying at up to Mach 12 and a spacecraft capable of Mach 25.

Rotorcraft

The unique qualities of conventional helicopters ensure that the vehicle has a continuing future for specialised tasks. However, its disadvantages of high initial and operating costs, noise, vibration, low speed and a poor safety record will prevent it from successfully competing with fixed-wing passenger aircraft over distances much greater than 150 miles. In the USA, the largest user of helicopters, attention has turned to the development of tilt-rotor aircraft that use two wingtip-mounted rotors to achieve vertical take off and landing, and at a safe altitude, tilt the rotors forward to become a turboprop powered wingborne vehicle, capable of greater speed and better fuel economy than conventional helicopters. The most advanced tilt-rotor project is the V22 Osprey, which is the subject of a $1700 million contract placed with the Bell and Boeing Vertol companies by the Department of Defense, with the Marine Corps being the most interested of the defense forces.

By early 1996, Bell Boeing were close to proceeding with the development of an 11 passenger civil tilt-rotor, the D600, which would replace medium size two-engined helicopters. It is likely that the new aircraft will weigh slightly less than 14 000 pounds (6350 kg), will have a cruise speed of 250 knots and a range of 1000 nautical miles. Aerodynamic (particularly stability) and mechanical complexities of the proposed tilt-rotor vehicle suggests that there would be formidable airworthiness problems to be solved before certification for civil use, and that operating costs, although greatly reduced from those of conventional helicopters, would be approximately 30% greater than those of a normal fixed-wing aircraft of similar performance.

Engines for subsonic flight

Engines being developed promise decreased fuel consumption, improved power-weight ratios, longer on-the-wing and total life, and lower cost per thrust/hour. These improvements are gained by the use of better materials enabling operating temperatures to be safely increased, and by improved gas flow and thermodynamics, but at the cost of increased complexity. Improved fuel economy will most probably be obtained by increasing bypass ratios of engines to a maximum figure of 10 : 1 without adding a gearbox – for both narrow body and wide body airplanes.

A search for reduced fuel consumptions has caused some engine manufacturers to seek to develop propfan engines that at first sight are a return to the obsolescent turbo-prop. Although the engine designs have much in common, they vary as to whether the propfans are ducted or non-ducted, whether they are driven directly or via gears and whether they are single row or double row with contra-rotation provided. Foremost among claims made for the new engines are that propfan technology will increase propulsive efficiency from the 60–65% achieved by high by-pass ratio turbo-fans to 75–85%, providing fuel consumptions that are improved by similar percentages.

Airworthiness authority experience with the propellers of now obselete large piston and turboprop-powered aircraft, will cause them to demand a full fault analysis and blade containment program and the UK Civil Aviation Authority (CAA) has stated:

'Extraordinary standards of engineering will be required for the new propfans.

'The challenge ... is that the propeller and gearbox and engine, must achieve a fatal debris-release rate from the start, not worse than that achieved by turboprops after 270 million hours, in spite of radical new technology.

'Certification standards will be "appropriately severe" and the cost could be "significant".'

The view of the CAA is that a propfan should be certificated as a turbo-prop because it is a 'rotating bladed device for which containment of a released blade is not provided ... with all that entails, for example conservatism in stress levels' (see Figure 12.4).

In order to achieve targeted safety levels, the CAA has calculated that the propfans will have to be three times less likely to lose blades than turboprops if they are to match the high safety standards of current and future turbofans. Airworthiness targets for catastrophic failures from *all* sources, is 10 per 100 million hours, of which one quarter are allocated to the power plant for certification purposes and the achieved (historical) rate of such failures for turboprop aircraft is 1.1 fatal releases per 100 million hours, i.e. within the required limits, whether caused by propeller, engine or gear box.

Flight test and development programs for the new engines will be extremely important and airlines and airworthiness authorities wait to see if claims made for the prop-fan will be met.

Engines for supersonic flight

Developed straight jet, i.e. non-fanned, gas turbine engines may be used for future manned supersonic aircraft that are to cruise within the earth's atmosphere at speeds up to Mach 3. The problems involved in the design, manufacture and operation of these engines are well understood, with relatively poor fuel economy and excessive noise being the only serious problems.

The Rolls Royce Olympus engines used in Concorde are old technology, being a developed version of military engines that are now more than 40 years old, but although using simple mechanical and analogue control systems, the Olympus has proved simple to operate and remarkably reliable. The in-flight shutdown rate of the Olympus is only 0.2 per 1000 hours which is not markedly inferior to that of some large subsonic engines, and it seems unlikely that further development of this type of engine would lead to any new safety problems, provided that conservative design and operating practices are observed.

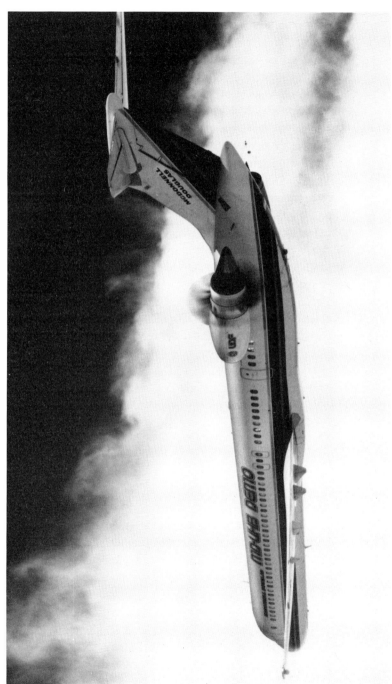

Figure 12.4 McDonnell Douglas MD-80 fitted with General Electric ultra-high bypass engines (*Courtesy of McDonnell Douglas*).

Engines for hypersonic flight

Problems of control and safety with rocket engines and storage and supply of rocket fuels seem almost certain to ensure civil passenger aircraft will be limited to operation within the earth's atmosphere for the foreseeable future, and air breathing engines are therefore likely to be used. Supersonic combustion ram-jets (scramjets) may be the chosen powerplants, and cruise speeds as high as Mach 7 are possible. Although the theory of scramjets is understood, there is as yet no experience in their operation, and it is possible that certification and safety problems would be encountered in introducing them into civil aviation.

There is some speculation that ceramic materials could be used in this type of engine but little is known about the service characteristics and safe lives of such materials as yet. Airworthiness authorities will be extremely cautious about giving approval to such radically new engines, and it is probable that this new technology will first be used in military aviation.

A new type of engine (named Swallow), proposed by Rolls Royce for a British concept of an aero-space plane, is a double function powerplant enabling air breathing propulsion through the atmosphere, and rocket propulsion as the outer limits of the Earth's atmosphere is reached when the vehicle is travelling at speeds between Mach 5 and Mach 7. The engine uses cryogenic fuels using liquid hydrogen and oxygen through-out the flight cycle, but its unique capability to burn atmospheric oxygen instead of on-board oxygen during climb to orbit would greatly reduce its launch weight.

Aircraft fuels

Increasing costs and postulated future shortages of fossil fuels, have caused attention to be directed at the possibility of using other fuels in civil air transport aircraft with cryogenic fuels such as liquid hydrogen (LH2) and liquid natural gas (methane – LNG) being considered. Weight reductions of at least 10% are possible by use of these readily available fuels, but difficult problems could arise in storage at the very low temperatures required, minus 423°F for LH2, and minus 258°F for LNG. Aircraft fuel tanks would have to be large volume and be extremely well insulated, but it might be possible to use the very low temperatures necessary to cool engine parts and for cabin air conditioning, so offsetting some disadvantages. The consequences for air safety in using these fuels have not yet been explored but they are inherently more safe than gasoline/kerosene and it is possible that no significant problems would arise.

Nuclear fuel has long been a goal of aerospace engineers wanting to exploit the enormous amounts of energy contained in a small mass of fuel. Difficulties encountered in shielding aircraft occupants from radiation, and the serious consequences for the general population should a nuclear-fuelled aircraft crash in a populated area, have prevented the possibility from being seriously explored as yet. It is probable that this type of engine and fuel would be used for military space missions rather than for passenger aircraft, but history has a way of showing that military inventions affect civil life at a later stage of their development.

Noise/emissions

A serious inhibition to the growth of civil air transport, is environmental nuisance caused by aircraft noise/emissions, and airline operations are adversely affected by international, national and local regulations imposed in attempts to limit the nuisance. Utilisation of extremely expensive aircraft equipment is reduced by airport curfews, extra costs are incurred when aircraft are delayed to avoid arrival at a curfewed airport and/or payloads are reduced in order to meet noise limits set by airport authorities. Great efforts have been made to reduce aircraft noise, and remarkable success has been achieved, with aircraft three times as heavy and powerful as the first jet powered aircraft, being only one-third as noisy when measured by the perceived noise decibel (PNDb) method.

Public opinion has not been satisfied by these efforts, and further attempts may be made by creating 'anti-noise' that is out of phase with the original noise thereby cancelling it, and this technique may be applied to new prop-fan engines which are likely to be very noisy. The first aviation application of 'anti-noise' techniques is in flying helmets for helicopter pilots. A future possible application would be to use anti-noise to stabilise some flow conditions, and perhaps control wing 'flutter', avoid engine surges and control aerodynamic flow – all important to airsafety. It is possible to use loudspeaker noise to induce wing vibrations to counteract wing flutter, to end a surge induced engine stall and to use sound to prevent major flow breakdown and/or oscillation.

The most cleanly burned hydrocarbons produce carbon dioxide and water as products of combustion, and carbon dioxide is a greenhouse gas with a residence time of about 100 years. Water vapour also has a greenhouse effect, but its residence time is far shorter, from a few days at sea level up to six months in the lower stratosphere Although aircraft contribute only 3% to the greenhouse effect they are very visible targets for environmental enforcement. Gas turbine engines with double annular combustors significantly reduce nitrogen oxide (NO x) and hydrocarbon emissions, and airlines are under regulatory pressure to use lower noise airplanes with less polluting engines.

Aircraft systems

Interest is being expressed in removing hydraulic, pneumatic and some engine-driven systems from aircraft, and using electrical power for the operation of *all* systems including primary flight controls, landing gear, flaps and spoilers etc. It is probable that an all-electric generation of airliners would use fibre-optic signalling between systems and components of systems, in efforts to reduce electro-magnetic interference and because of the much faster signalling rates available. Advantages claimed for all-electric aircraft are reduced weight, simplicity, cleanliness and absence of corrosive and flammable substances, but there may be resistance to the proposal from airline engineers who find it easier to trace a hydraulic leak than electro-magnetic interference between two unrelated but close proximity electrical systems. All-electric systems would need careful design if they are to provide the same 'redundancy' as conventional aircraft systems.

A simple but safety-effective system that should be provided, is a video monitoring system to enable the flight crew to see the aircraft cabin, wing tips, tail surfaces, landing gear and engines while in flight or on the ground. Such a system would have been invaluable in a number of past incidents, accidents and hijackings when pilots have been deprived of information that would have enabled them to better contend with abnormal circumstances. In recent years there have been occasions when pilots' wrong identification of the source of engine fire warnings resulted in a tragic loss of lives. It would be possible to design a system that would present a video image to the pilots only when image comparison shows that an abnormal situation has arisen.

Materials

Carbon fibre reinforced plastic (CFRP) composite materials make up about 10% of the structural weight of the B777, and provide weight reduction, resistance to fatigue and corrosion, resistance to impact damage and reduction in need for structural repairs. Materials used are resistant to standard aircraft liquids such as hydraulic fluid, jet fuel, and de-icing fluid with service proven finishes applied for protection against moisture, ultraviolet light and erosion damage. Less well understood is the performance of some of the new materials with regard to electromagnetic compatibility, resistance to lightning strikes and a tendency of epoxy resin matrices to slowly absorb and retain water. The FAA and other regulatory authorities have been active in devising acceptable inspection and repair programs for aircraft already certificated for civil use, and lessons will be learned from existing military aircraft programs where wings and fuselage are formed from these new materials. In the military F22 program, Lockheed used a total of 35% composites and achieved a weight reduction of 25%.

Research and development in countries with major active aircraft construction programs are bringing a quickening of this revolution in materials used in airplanes and engines. In the USA, NASA invested $142 million over a six-year period, awarding contracts to Boeing, McDonnell Douglas, nine other firms and four universities, and the results are beginning to show. A NASA sponsored Advanced Composite Transport (ACT) Program has objectives that include reducing manufacturing costs by 25%. In a similar program, Airbus is studying a composite fuselage and wing for aircraft that may be in service in the year 2010.

Engine designers striving for higher and higher thrust-to-weight ratios, and lower operating costs have hit a barrier where superalloys operating at 80% of melt point require large quantities of cooling air to be bled from the compressor. When ceramics or non-homogenous composite materials with higher temperature tolerances can be used, great gains in efficiency will result from abandoning wasteful use of cooling air. Many people believe that future engines will have a composite compressor and ceramic turbine!

Automation in the cockpit

Aircraft and systems designers have shown that almost any system and procedure can operate with little or no intervention from man, but a problem yet to be solved is the best way of mixing the best features and capabilities of man and machine, for no one has yet suggested that airliners should be completely automated and operated without flight crew. Aircraft systems, displays and procedures coming into use during the late

1990s are an attempt to find a best mix, but doubts are expressed about some of the solutions adopted.

The problem of deciding what information to display to the crew and what in-flight interventions they are permitted to make, was increased by introducing a minimum crew complement of only two pilots, and it is probable that future aviation historians may not be able to decide which came first, the decision to automate or the decision to reduce crew complement.

In 1986, when reduced crew complement and automation were seen as related development, the following reactions were evident.

At a Flight Safety Foundation Seminar:

- Only man has a potentially unlimited capacity to receive and process data and information, and to decide logically the priorities and possible ways to fight abnormalities, even in unusual unforeseeable situations and conditions
- Many environmental situations, among them everchanging weather, will never be adequately programed for use in computers
- Automation decreases monotonous workload
- The degree of automation should be structured to achieve an optimum workload and arousal level for the crew
- Automation should leave the decision of its usage to the pilot (or management)
- Automation failures should not lead to additional workload for the crew

Surveys of airline pilot opinion conducted by research establishments in the USA and United Kingdom, and reported to the Flight Safety Foundation and Royal Aeronautical Society show pilots believe that:

- Flying skills are degraded by automation
- There is over reliance on automation
- Automation on flight decks is a 'good thing' and reduces crew fatigue
- Most pilots enjoy flying automated aircraft

Analyses of incident and accident reports show that on occasions there are mismatches between the role of systems and pilot perceptions of that role, and accidents have occurred because the pilot either failed to understand what the system was programed to do, or failed to intervene when the automatic system was improperly programed, or was malfunctioning.

In general, pilots like the new flight instrument displays and autopilots and believe they have a favourable effect on safety. They also believe that the flight management system's (FMS) prime function is to improve the economics of air transport and is not user friendly, the controls are unduly complex and the work load involved in reprograming when the intended flight plan has to be modified is excessive – particularly when flying in terminal areas.

Perhaps in recognition of pilot response to FMS, NASA at Ames Research Center was developing in 1995, a new approach to computer-based learning in flight deck automation training. A key feature of a training system designed to help pilots make the transition to the modern glass cockpit aircraft, is the use of a Macintosh laptop computer and a moderate to high fidelity software emulation of a typical FMS.

Reliable information suggests that IATA is concerned about accidents involving highly automated aircraft. An internally circulated report shows that an investigation team visited all major airplane manu-facturers to express concern. Comments made in their report include:

- A new class of 'design induced' accidents are occurring in highly automated aircraft
- These accidents are associated with design features that adversely affect the interface between the aircraft, its systems and the pilot
- This interference may also involve the removal of tactile, aural or visible feedback that was usually present in earlier systems
- 'Inhibits' intended to safeguard the airplane, can effectively inhibit perfectly correct, normal or instinctive actions by the crew
- Inadvertent selection of auto-modes, leave pilots without adequate control authority
- Design philosophies that can confuse the pilot, remove reasonable decision-making ability or increase workload are counter productive
- Permutations and combinations of multiple and partial failures appear not to have received adequate recognition
- Further research into man-machine interface is needed to achieve a better balance
- FMS is by design open to human error
- It is essential to reverse this trend of design-induced accidents

Because IATA airlines are 'customers', manufacturers are sure to pay attention to these strongly worded comments.

Cockpit automation (B747–400)

The basic B747 (see Figure 12.5) was introduced into airline service in 1970 and the latest version on offer to airlines is the B747 – 400 (see

Figure 12.5 Boeing B747–100 – 3 crew flight deck (*Courtesy of Boeing*).

Figure 12.6 Boeing B747–400 – 2 crew flight deck (*Courtesy of Boeing*).

Figure 12.6) which reduces operating costs by approximately 15% over early models. A significant change from earlier models was the reduction of flight crew from three to two with the role of the flight engineer being automated and/or simplified and all controls fitted into the pilots' overhead panel eliminating the flight engineer station. Boeing states that the design has reduced the number of cockpit lights, gauges, and switches from more than 970 in the basic B747 to only 365 in the –400, constituting a triumph of design and engineering skills.

There is increasing agreement in civil aviation, that good design of automatic systems can relieve men from the monotonous chore of monitoring systems and so free them for higher cognitive tasks that are as yet beyond the capability of machines, and the B747–400 is a good example of application of these principles.

Computers – artificial intelligence (AI)

Much information is presented to flight crews and the use of advanced sensors, computer generated displays, and multi function displays, greatly increase the amount of information available for decision taking. As human capacity for absorbing, digesting and making decisions cannot be increased, it is possible that AI will be introduced in military airplane cockpits and later in civil air transport aircraft. It is envisaged that AI computers used in 'expert' systems would integrate information, set priorities, filter, and then communicate the most significant information in the context of the current situation. Because extraneous information would be suppressed, pilots could easily comprehend critical factors and choose from a range of alternative courses of action presented by the AI computer. In its ultimate form, the AI system could execute the 'best' alternative available with, presumably, the human pilot having power of veto.

Proponents of the use of AI systems in airline aircraft point to evidence that accidents have occurred when sufficient information was available to the pilot to prevent the happening, but the pilot failed to use that information for some reason, such as being distracted by a minor event or by reason of excessive workload. They argue that AI systems could not become overloaded, be distracted or be adversely affected by fatigue or stress. Monitoring is an activity at which AI systems would excel, and when a crew input is required the system would alert pilots. Possible uses for AI include:

- Monitoring the status of aircraft systems
- Diagnosing and predicting faults

- Warning of emergencies
- Recommending responses to emergencies
- Navigation/route planning
- Fuel management
- Contingency planning
- Avoidance of severe weather and collision risks

In 1988, Lockheed Georgia won a NASA research contract to provide airliner crews with flightpath advice for unscheduled diversions and devised a system called 'Diverter'. Early demonstrations had 'Diverter' responding to a severe weather message with a recommendation to divert and suggesting a diversion location based on factors such as weather, safety, economy, schedule and airport facilities. A second demonstration was to interface 'Diverter' with a NASA Langley systems status project called 'Faultfinder'. Using 'Faultfinder' the system could, for example, detect a pressurisation failure over the Rocky Mountains requiring an immediate descent and 'Diverter' would then suggest a route through the mountain range. By 1994, European researchers were working along similar lines again concentrating on assisting pilots when an unscheduled diversion had to be made. Early work on the program suggested that the AI computer and human pilots generally make the same decisions but the computer is faster.

There must be doubt that such systems are necessary, when a data base in a typical flight management system can store operational information for *all* usable airports along an airline's route system, and navigation is more accurate and less effort-intensive. The FAA has already warned of difficulties that certification of AI systems will present, and is including a dedicated section on AI in its Digital Systems Verification Handbook.

An interesting development is attention being given to 'neural network' computers that function more like a human brain than digital computers, and have an ability to learn from mistakes. No doubt these will be even more difficult to certificate! The first application of 'neural network' type software may be in military airplanes where the software could compensate for battle damage to flight control systems by comparing what is actually happening to the aircraft with data on how the aircraft should fly. If there is a mismatch, the software using basic aeronautical equations attempts to modify the aircraft's control laws.

Speech recognition systems

A different approach to the problem of too-high pilot workload is aimed at augmenting the ways in which pilots can interface with aircraft systems by providing voice control of systems such as radio frequency selection, inputs to displays and the entering of data. Research has been performed on the difficult task of designing an airborne word recognition system that can perform fault-free in the noisy environment of typical cockpits, and with differing voices of human pilots. First applications for such systems are military, but eventual adoption in civil airline aircraft is almost certain.

The basic rationale for introducing such a system is that the eyes and hands of pilots are overloaded, and this perception has already led to synthetically produced voices being used for warnings such as 'pull up, pull up' in ground proximity warning systems, and to give other aural warnings at times of high cockpit workload. It seems possible that a fully developed voice recognition system would increase the performance of a crew if for example, the captain could 'tell' a system to perform some task, instead of commanding the copilot to perform it. In these circumstances the possibility of increasing the copilot's capacity to monitor all safety critical actions performed by the captain or by automated systems would be increased.

Canada's National Research Council began helicopter flight trials of a speech recognition system in mid 1995, and first results indicate that a 99.6% success rate is achievable in this difficult environment.

Thought control of aircraft systems by the human computer

The Workload and Ergonomics Branch of the USA Air Force Aerospace Medical Research Laboratory has for many years conducted research into the possibility of using the human brain's electrical signals to the body, to activate aircraft systems directly. These signals can be measured and if detected by very sensitive sensors, could perhaps be amplified and be passed by super conductors to systems to be activated.

By early 1995, test pilots at the Wright Patterson Air Force Base were making simple control inputs into a flight simulator using only the power of thought after learning to vary the intensity of their brain responses to stimulation by pulsed fluorescent lights. On the simulator, heightened intensities cause the simulated aircraft to bank to the right, depressed intensities cause it to bank to the left. It is interesting that 'the subjects are unable to say exactly what they learn to do but it's like learning to walk, after a while they no longer have to think about what they are doing', according to the director of biocybernetics at the base. As one of the neuro-physiologists in the program said, 'One of the main advantages of the techniques is to give pilots a better chance of managing the increasing amount of data that is imported to combat aircraft cockpits. If your hands are busy with other tasks, it would be nice to be able to think; "I would like to see that other screen now" or to choose items from an on-screen menu'.

It may never be possible or desirable to use thought control to actually fly an aircraft, but the research program shows some promise of revealing methods of monitoring brain activity so as to be able to measure pilot fatigue, work overload and, in military aviation, the onset of g induced loss of consciousness. The program is of great interest to all who are interested in the problems of providing better interfaces between man and machine.

Information technology and computers

The rate of change in aerospace information technology is phenomenal, and has been essential to progress made in recent design, construction and operation of aircraft.

The first modern computer was invented in 1946, named ENIAC and located at the University of Pennsylvania USA. It weighed 39 tons, stood two storeys high, covered 15 000 square feet, employed thousands of valves, miles of wiring, thousands of soldered joints and needed many kilowatts of power and air conditioning to function. That same computing power can now be purchased for a thousand dollars, and be carried away in one hand by the purchaser.

In the aerospace manufacturing industry, computers are available at all levels of the design and engineering departments, and Computer Aided Design (CAD) has become a standard working method. Computers are linked together permitting designer-engineers to be physically separated by thousands of miles when working on joint international projects, and yet be able to access the same data base and communicate with each other without hindrance. Computers handle more than 40 million instructions per second, and more powerful computing power is available if required for special purposes, with so-called super computers being capable of performing approximately one billion calculations per second. It is forecast that computers will be capable of 10 trillion operations per second by early in the 21st century, using parallel processing and optical computing techniques.

Virtual reality has become a useful tool in design for the task of maintenance evaluation in complex machinery spaces, and provides better visualisation of complex designs, earlier design verification, designing for ease of assembly and maintenance and the training of maintenance engineers. It is used as a substitute for physical pre-assembly of engines, and their attachment to airplanes, and is regarded as an advanced tool for the design, prototyping and implementation of human-system interfaces, by moving and manipulating objects using a standard mouse and spaceball 6-degree of freedom (X, Y, Z, roll, pitch, yaw) strain-gauge hand controller.

Safety is enhanced by these new design and manufacturing methods, and there is much less risk of an aircraft showing unexpected and undesirable flying qualities in test flights or in airline operation than was formerly the case. Almost all of the flight envelope is investigated by flight simulators before the aircraft makes a first flight, and required changes are incorporated at an early stage of a project. Pilots from customer airlines 'fly' a new aircraft type before the design is frozen with their ideas and criticisms being taken into account, and built into the aircraft with good results for all concerned. There is much less chance than ever before of projects failing because of unforeseen technical shortcomings, and the whole air transport industry has benefitted from the information technology/computer revolution.

A hostile environment for computers

Computers need protection from extremes of temperature, humidity, vibration, power surges, lightning strikes, interruptions of power supplies, electro-magnetic interference and static electricity, and great efforts are made to provide computer installations with an ideal environment. If a computer installation is ground-based, this task is simple although expensive, but even in near ideal environmental conditions, computer defects or 'glitches' continue to occur, as shown by frequent occasions when computer systems vital to the operation of banks, hospitals, armed forces, air traffic control and airlines are reported to be 'down'.

When a computer system essential for safety is used in aircraft, providing a near ideal environment is much more difficult. Aircraft structures vibrate and suffer g forces, outside air temperatures vary from plus 40°C to minus 40°C, humidity varies from near 100% on the ground at sea level in the tropics, to only 16% or less at the end of a long flight. Shut-down of an engine loses half of the installed generating capacity in a two-engine aircraft, reducing cooling airflows, and an aircraft environment is generally hostile to computers. Although power supplies are protected, cables are shielded, aircraft structures are bonded against lightning strikes and static discharges, computers are installed on shock absorbing mountings, air conditioning is provided and special cooling air supplies are available even in emergency conditions – the airborne computer has a more adverse operating environment than its sister on the ground, although its continuing function may be essential to safe flight.

Electro-magnetic interference is defined as undesirable voltages or currents that adversely affect a system, and the effects may become evident as static in audio receivers, inaccuracies in instrument indications or 'herring bone' patterns in video projections. Some sources known to emit energy creating interference are: fluorescent lights, radio and radar transmitters, power lines, window heat controllers, induction motors, switching and light dimming circuits, pulsed high frequency outputs and lightning. This energy can reach a circuit or system by conduction through electrically conductive paths such as circuit wiring or aircraft metallic structures, or electro-magnetic field radiation through electrically non-conductive paths such as air and fibre glass. Electro-magnetic interference may be reduced by techniques such as:

■ Suppression at source
■ Shielding by metallic housing or twisting/shielding noisy wires
■ Reduced noise coupling – by separating power leads, using filters, twisting/shielding noisy wires – perhaps by using co-axial cables
■ Increasing the susceptibility threshold of receiver

Reports have been made of other adverse circumstances that may be peculiar to airborne electronic equipment including computers, and it is known that aircraft and system manufacturers are aware of the need to take special precautions to minimise effects on reliability and performance.

There have been instances of pilots reporting what British Airways (BA) has called random memory initialisation (RMI) of cockpit computers. The faults are most common in the early part of a flight after the aircraft has been on the ground for some time, and has been experienced in *all* aircraft types operated by BA. The theory is that chips are retaining instructions that should have been overridden by pilots' latest entries to the flight management system computer. Potentially dangerous examples of this are when the system changes what the pilot has entered into the 'height acquire' function of the autopilot. Pilots report that they enter the required height value and see it displayed in the mode control display only to have the data change to the earlier value and the display change to that value. When this has not been noticed by pilots, aircraft have climbed/descended to a height not authorised by air traffic control and introduced the risk of an air collision.

In another manifestation of this problem, the 'height acquire' display settles one flight level above or below that selected, and BA names this a transient electromagnetic condition (TMC) which it claims has now been cleared by modification. TMC can be induced by current changes and is a well-known electronic phenomena. Power surges or other forms of TMC can occur under well recognised conditions such as changeover from ground to aircraft power, lightning strikes, and proximity to radar transmitters etc., but there have been some where the cause was not obvious.

Early manifestations of these and similar random short duration failures of complex systems in airplanes, were studied in the 1980s by the German pilot association which found that faults were caused by Alpha particles and cosmic rays destroying information stored on chips. These 'soft fails' were short term because the chips are interrogated and freshly charged at intervals of a few micro seconds. The study showed that the failure rate increases dramatically with increasing altitude from sea level, to a factor of 150 at 33 000 feet.

Other causes of short term failures in aircraft equipment are the operation of lap-top computers, and some types of portable cellular telephones carried aboard airliners by passengers. The NASA Aviation Safety Reporting System (ASRS) revealed that in the period between April 1988 and December 1992 there were 24 reported instances of passenger operated electronic devices causing interference to aircraft flight controls, flight instruments or communication and navigation equipment. Pilots and airlines want the use of such equipment to be banned in flight, but regulatory authorites seem inclined to compromise and to allow their use in cruising flight.

Future operational environments

Faced with widely varying regional air navigation plans that in some instances would never be implemented, and which in even the best region fell widely short of current requirements and/or were incapable of expansion to meet future needs, ICAO in 1983 appointed a committee to study and report upon Future Air Navigation Systems (FANS). Its landmark report was adopted nine years later to become known as the ICAO Communications, Navigation, Surveillance and Air Traffic Management systems concept – abbreviated to CNS/ATM, and a global coordinated plan for transition to the new concept was agreed in 1993. The use of satellite technology was seen as the only viable solution meeting the needs of civil aviation on a cost-effective global basis, and the system is a mix of satellite and terrestrial systems to achieve an optimum result.

When the new CNS/ATM systems enter operation, communication with aircraft for both voice and data (except for polar regions) will be by direct aircraft to satellite links and thence to the air traffic control centre via a satellite ground earth station and ground-ground communications network. In terminal areas where line of sight communications are not a problem, VHF voice augmented with data link and SSR (Secondary Surveillance Radar) Mode S data link will be used. The satellite communications system will ultimately eliminate the need for high frequency (HF) voice, although HF may continue to be used over polar and other specific areas in the short term.

The global navigation satellite system (GNSS) will eventually be suitable as a sole means of navigation, and replace the current large variety of short-range navigation aids. The United States' global positioning system (GPS), the Russian Federation's global orbiting navigation system (GLONASS) and the International Maritime Organisation's Inmarsat navigation transponder equipped satellites will be used. GPS has 21 operational and three-in-orbit spares circling the earth twice per day in six planes in circular 20 000 km orbits.

Surveillance on a global scale will be by automatic dependent surveillance (ADS) with aircraft automatically transmitting GNSS derived position, speed and other data, including environmental data, via a communication link to the ATC center where aircraft position will then appear on an electronic display. An important feature of ADS is that the downlinked data will provide much more data than present radar systems. Secondary Surveillance Radar (SSR) augmented with Mode S and improved antennae will be used in terminal areas and high density airspace.

With the implementation of CNS systems complete, states will have the necessary basic tools to provide ATM that can handle foreseeable traffic densities and complexities. In low density areas the ATM will provide safety and regularity without imposing heavy economic burdens on the service provider. In high density areas it will be necessary to implement powerful automated systems to respond to ever increasing demand, but it must be possible to equip aircraft used internationally with a *single* set of avionics that is usable everywhere, limiting variations of equipment to the ground part of the system.

When the systems are fully implemented, there will be nowhere on earth where aeronautical communications, navigation, surveillance and air traffic management are not available and this will be a great bonus for air safety. The first ICAO region to fully implement CNS/ATM may be the South Pacific with a target date of January 2000, and it is planned to have fully implemented the system in North Atlantic airspace by the year 2005.

Somewhat surprisingly, doubts have been expressed about the legal status of GNSS usage in international airspace, and in 1996 the ICAO Legal Committee had that item added to its agenda with a view to establishing an authoritative ruling that will remove all doubts. With less than four years before the first region is scheduled to have GNSS fully implemented, the Legal Committee will have to move with unusual speed to resolve this issue.

A USA perspective

Writing in the November 1995 edition of *Air & Space*, a journal of the Smithsonian Institution, David R. Hinson, Administrator of the FAA titled his piece: 'Goodbye Yellow Brick Road', likening 1995s air routes to the convoluted path in the Wizard of Oz. Mr Hinson sustained his choice of title by pointing out that the aerial highway between Los Angeles and Chicago is relatively straight through California and Nevada, turns northeast through Utah and Wyoming to avoid traffic outbound from Denver, makes a sharp righthand turn and then winds through South Dakota, Iowa and southern Wisconsin before descending into Chicago O'Hare. Along the way controllers direct pilots to maintain certain speeds, altitudes and safe distances from other aircraft. These techniques deny flexibility to pilots to take advantage of available airplane performance, shifting winds or weather conditions, and with more airplanes filling the skies these operating restrictions cause delays that cost airlines and their passengers an estimated $3 billion per year.

Mr Hinson went on to state that a convergence of satellites, computers and digital communications technology offer a revolution in air traffic management, a concept called 'free flight'. In its ultimate form 'free flight' would let pilots fly whatever route is best under existing conditions, reducing flight time and avoiding weather or turbulence, but pilots would still be required to file a flight plan. Controllers would continue to play a vital role in air traffic management, modifying filed flight plans if they posed a safety hazard, and receiving automated altitude, speed, position and heading information from aircraft in flight. They would only intervene if computers predict that an aircraft's 'alert zone' – a chunk of airspace shaped like a hockey puck – would violate the same zone around another airplane.

Airlines are enthusiastic about the idea, with American Airlines estimating that a reduction of only 2 minutes per flight would save $40 million annually, and United Air Lines stating that full implentation of 'free flight' would cut its yearly operating costs by $600 million to $1 billion, with another $800 million saved by increased productivity. The FAA is committed to implementing the concept but only after ensuring that safety is maintained, every advanced technology required for free flight works as it should, and after taking an evolutionary approach to implementation.

Important building blocks to be used are, satellites of the Global Positioning System (GPS), fast and error-free data-link communications between cockpit and ground, highly reliable advanced automation systems in air traffic control facilities and next generation Traffic Collision and Avoidance Systems (TCAS) aboard aircraft. Some of these items are closer to introduction than others and 'free flight' and the current structure of air traffic control will exist side by side into the next century in USA airspace. In upper airspace, limited availability of free flight already exists (1996) and data link communication trials are taking place over the Pacific.

Great pressures exist to ensure that the FAA goes ahead with 'free flight', not least because the number of passengers and airplanes will increase by more than 20% over the next decade, and the number of US airline operations is forecast to increase from 38.3 million in 1994 to 44.1 million in 2000. General aviation, which creates many more flights than airlines, will also grow but it may prove difficult to integrate general aviation flights into the new system because of the expensive airplane systems needed for participation. Human aspects to be considered are changed roles and responsibilities of pilots and controllers, and these will require delicate and careful negotiation before real progress is made.

Maritime initiative – Inmarsat

The International Maritime Organisation's (IMO) International Maritime Satellite (Inmarsat) organisation owned by telecommunications organisations from more than 75 countries was the first international agency to introduce communication satellites. Its four operational geostationary aero satellites (see Figure 12.7) plus in-orbit spares provide two-way voice, fax and data services for aircraft operating virtually anywhere in the world, except extreme polar regions. The Inmarsat–3 satellites will carry an onboard navigation package to allow broadcast of augmentation and integrity data, related to the current GPS and GLONASS satellite and navigation systems.

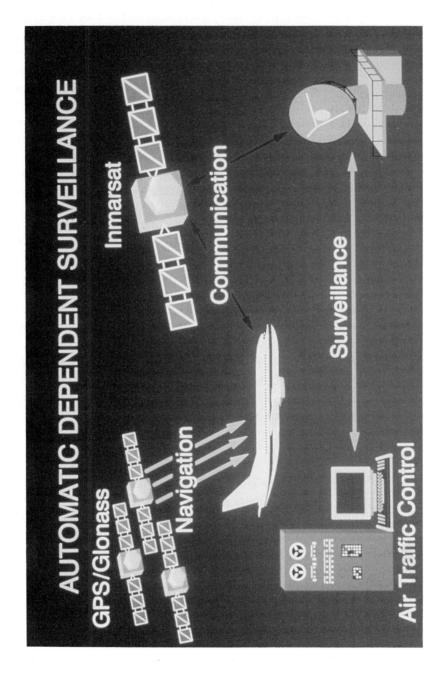

Figure 12.7 Inmarsat satellite (*Courtesy of Inmarsat*).

Global Positioning System (GPS)

The satellite based navigation system GPS, developed for the USA military and with accuracy deliberately reduced for civil aviation, will provide commercial users with horizontal position to within 100 meters and vertical position to within 160 meters plus velocity within 0.2 knots and time within 100 nano-seconds. The system works by triangulation from three satellites and range from a fourth satellite providing aircraft position in three dimensions plus time. Twenty-four satellites in orbits of 10 500 nautical miles are required to provide 24 hour worldwide coverage, and operation with all 24 satellites could provide automatic dependent surveillance that would enable aircraft to automatically transmit data from on-board navigation systems to ground stations via satellite datalink, thus permitting automated tactical air traffic control in all airspace. The UK is reported as being ready to approve the use of GPS as a prime navigation system on North Atlantic routes when the ICAO North Atlantic systems planning group meets in September 1996.

Glonass

Three Russian Global Orbiting Navigation System Satellites equally spaced in stationary orbit, plus three in 12 hour elliptical orbit close to polar orbit can provide communications and automatic independent surveillance on a global scale. One satellite in elliptical orbit with apogee in the southern hemisphere, would be assigned to service flights in the Antarctic region, and could provide 17 hours of service per day, reckoned to be adequate for that area. The other two satellites in elliptical orbit with apogees in the northern hemisphere, could provide round the clock services for the Arctic region and partially for central latitudes. Three of the satellites in geostationary orbit could provide round the clock communications and dependent surveillance from 70° north latitude to 70° south latitude. In the ICAO global navigation system, Glonass will be used with GPS and Inmarsat satellites to provide the CNS/ATM function.

Trials of a combined GPS/Glonass receiver conducted by the UK Civil Aviation Authority showed that combined availability never fell below ten satellites, while GPS-only availability could fall as low as four, with the added advantage that Glonass signals are not deliberately degraded as are GPS. Most industry experts believe that absolute integrity can only be achieved with a minimum of six satellites in 'view'.

Data link

An operative example of a communication data link between air and ground is the Aircraft Communication and Reporting System (ACARS). Messages are sent from the airplane by VHF transceiver via a ground network to the airline ground operations base and vice versa. Flight crew may use a control unit keyboard to send a message, or it may be generated automatically by ACARS and one of its interfacing airplane systems. ACARS relieves the crew of sending many routine voice radio messages by downlinking preformatted messages at specific times of the flight with each message being compressed to take about one second of air time to transmit. Ground personnel may request data at any time and the uses of ACARS are almost endless varying with the airline. Items include:

■ Dispatch and weather items
■ Engine performance
■ Fuel status
■ Weather services
■ Automatic Terminal Information Service
■ Flight status (delays, diversions etc.)
■ Passenger services
■ Maintenance items

Airlines investing in comprehensive airplane data recording of hundreds of operating parameters, obtain the best return from ACARS as their maintenance departments can monitor the detail of flight operations on a continuous basis. In North America, the system is almost wholly VHF but 'phone patching techniques are used by some airlines making it possible for the ACARS system, or derivatives of it, to function on a worldwide basis, thereby transforming airline control of operations. Writing in the *ICAO Journal* of June 1995, a representative of the International Air Transport Association (IATA), stated with confidence that a HF data link network could play a role in the provision of air traffic services, with 95% availability when coverage is provided by two interlinked HF stations, and over 99% with coverage by three HF stations. He saw HF data link and satellite communications as being complementary.

When the ICAO Communications, Navigation, Surveillance/Air Traffic Management system CNS/ATM is implemented, satellites will permit VHF data link to be used world-wide so that operational information required to operate CNS/ATM is always available. This will provide enormous twin benefits of improved safety and reduced costs to all of the civil air transport industry.

Vertical separation between aircraft

Since the earliest days of aviation altitude has been indicated by altimeters using aneroid capsules to measure changes of pressure. The instruments are calibrated to indicate altitude and not pressure by using a generalised relationship between altitude and pressure, but because that relationship is not constant the calibration of altimeters is a complicated task. The deficiencies of this method of measuring altitude are well known, and includes the need to compensate for instrument errors, aircraft pitot/static systems, temperature and the Mach effect etc.

For a pilot to know vertical clearance over the earth's surface, it is necessary to know the prevailing pressure at ground level, or the prevailing pressure at sea level together with a knowledge of the height of the ground over which the aircraft is flying. To have this information requires that the pilot navigates accurately and that considerable time and effort be expended by meteorological observers and air traffic controllers, placing large burdens on communications systems.

Radio altimeters are used in airline aircraft only to precisely measure height above ground level for special purposes such as autoland systems, and are rarely used at heights above 2500 feet. Radar altimeters were used in the 1950s/1960s while cruising at medium levels, for special purposes such as en-route pressure pattern flying, but increased air traffic put an end to that technique as it was wasteful of air space and made difficult the ATC task of ensuring separation between aircraft.

The use of 2000 feet vertical separation above flight level 290 (29 000 feet), compared with 1000 feet vertical separation at lower levels, was long seen by airlines and states as a restriction that should be removed, but ensuring safety with 1000 feet separation at higher levels proved difficult to achieve.

A proposal favoured by IFALPA was to retain the 29 000 feet demarcation between upper and lower airspace, with vertical separation below the demarcation level to be 1000 feet, but to use a gradually increasing vertical separation above that level based not on feet but on a value of atmospheric density. This system was discussed in civil aviation as early as 1969 when it was called the ICA system, but IFALPA policy supports an amended system called COS Lanes, for Constant Optimum Separation Lane system of vertical separation, claiming that only new software would be needed to implement the system and that costs could be recovered in three years. The IFALPA proposal has failed to find support.

In 1994 ICAO approved 1000 feet separation to be introduced on a regional basis, and the target date for the North Atlantic region is 1997/98, with European airspace following in 2001. The implementation program includes:

- Improved ground and airborne systems
- Arrangements for minimum aircraft system performance specifications (MASPS)
- Procedures for approval for operating in reduced vertical separation (VSM)
- Airspace and system performance monitoring
- Height monitoring of aircraft in level flight (necessary to verify that system safety is not compromised and an integral part of system performance monitoring).

The program when implemented will greatly increase the number of aircraft that can be accommodated in the upper airspace, but some operators of general aviation airplanes unable to meet equipment costs will find themselves restricted to flight levels lower than 285 (28 500 feet).

World meteorological services

Two world area forecast centres, WAFCs, at Washington and London England, provide global computer forecasts of significant weather that is facilitated by an increased number of air reports sent automatically from aircraft by air-ground data link. The first satellite broadcast services from Washington were in May 1995, and this new service provides states with reliable access to standardised high quality meteorological information to support accurate and timely flight planning for international air navigation. The US-provided satellite broadcast service covers central America and the Caribbean, North and South America, North Atlantic region, the Pacific and Eastern parts of the Asian regions. The rest of the world is covered by London. Data on upper winds and temperatures are in digital grid-point form, and in chart form, and together with significant weather forecasts, aerodrome forecasts and aviation routing weather reports, provide a complete service that is confidently expected to prove of great benefit to international air transport operations.

In the USA, automated surface observing systems are being installed by the National Oceanic and Atmospheric Administration (NOAA) and FAA (at key airports) and these are expected to greatly improve the reporting and forecasting of weather for aviation in the USA.

Airports

In developed countries, environmental restrictions make difficult the construction of new airports and increasing use is being made of landfill techniques to create new islands or peninsulas as at Kansai, Osaka, Japan and Hong Kong off the coast of China. The new island airport of Kansai sunk by 35 feet in only one year, and expensive new work has to be performed. This persuaded Japanese authorities to consider building a floating airfield for Tokyo probably located in Tokyo Bay, and in early 1996 a conglomerate of shipbuilders and steel makers were at work on the project with thirty sites under consideration. An experimental platform costing 7.5 billion Yen ($35.5 million) and measuring 1000 feet by 200 feet is planned for testing in early 1997. If the project goes forward it is possible that the main structure will be built from simple box floats located 2 miles offshore to be protected by breakwaters and anchored to the seabed by cables. It is expected that use of the best anti-corrosion materials and techniques would give a maintenance life of at least 100 years.

Advanced landing systems

Increased risks and disruption caused by low cloudbase and restricted visibility at airfields were early recognised, and before World War II, Major Jimmy Doolittle made 'blind' landings using gyroscopic flight instruments invented by Elmer Sperry, and in 1937 Lieutenant Crane made the first completely automatic landing. These were the first attempts to compensate for absence in cloud or fog of visual 'cues' used by pilots to position airplanes in space so as to make safe landings.

In the 1960s, British airlines developed automatic landing systems using failure-surviving triplex autopilots coupled electronically to lateral and vertical radio beams of ground-based Instrument Landing Systems (ILS), and augmented by radio altimeters for the landing flare. These systems are widely used and vary only with the level of redundancy installed, with triplex systems providing 'fail active' performance and duplex systems being 'fail passive'. Automatic landing systems achieved remarkable success, and by the end of 1995 there had been no fatal accident caused by them – perhaps because computers and machines do not need visual references and do not suffer from fatigue and stress caused to human pilots by the physical and mental demands made when flying solely by reference to flight instruments while in close proximity to the ground.

A major advantage from frequent use of auto-approach or autoland equipment, is that approaches to runways are consistently more accurate than manual approaches improving safety during landing phases of flights. Standard Category 1 (non-precision) ILS are useable by autoland systems in good weather. Airlines encourage pilots to carry out a high percentage of approaches to land by use of automatic systems, but accept that human pilots need to practice manual skills for use on occasions when air or ground components of the system are not available, or conditions are beyond the capabilities of the automatic systems for reasons of strong cross winds or a runway surface contaminated by ice and snow.

A USA NTSB study of 17 ILS approach accidents between 1970 and 1975, revealed that most occurred after the runway had been sighted and the pilot made a transition from flight by instruments to flight by visual references as the primary means of determining the descent path to the runway. A study conducted by NASA and Boeing showed almost 50% of jet transport accidents between 1959 and 1982 occurred during final approach and landing, with a high correlation between hand-flown approaches in low visibility and approach accidents.

Europe's JAAs-agreed operating limits/performance specifications for landings are:

CAT(egory)	Decision height (feet)		Runway visual range (meters)
1	200	(no approach lights)	1000
2	Higher than 100	(auto coupled)	300
3a	Less than 100	(fail passive)	200
3b	Less than 50	(fail operational)	125
3c	0	(fail operational)	75

NB: USA limits are different and permit use of approved HUDs in low weather minima.

By the early 1990s, after huge investments by airlines and airports, the best automatic landing systems enabled landings to be made with visibilities of only 50 meters and a zero cloud base. The only disadvantages were costs, a shortage of radio frequencies in the waveband and ILS systems not installed at some airports because of terrain or other physical characteristics. This caused ICAO to determine what precision landing aid should replace ILS, with the eventual decision being to install Microwave Landing systems (MLS), using a waveband that had plenty of frequencies available, could provide curved approaches and tolerate difficult sites such as Valdez in Alaska, which was one of the first airports equipped with the new precision landing aid.

A decision was made to 'protect' ILS until 1995, with an assumption that MLS would be widely installed and in use by that date. However, doubts about the need for MLS began to arise when the potential of using Global Navigation Satellite Systems (GNSS) derived spatial information to make precision landings, became apparent as early as 1992, and so the 'protection' date for ILS was moved to the year 2000.

By 1995, ICAO was faced with two distinct groups of states and user airlines, one group wished to stick with the decision to introduce MLS by the year 2000 and the other, led by the USA, wanted to wait and see what GNSS based systems could do. A compromise was adopted, to retain ILS 'for the foreseeable future', encourage limited implementation of MLS 'at locations where it is operationally required and economically beneficial', promote the development of a 'multi mode receiver' capable of using ILS, MLS and GNSS signals, and promote GNSS research and development. A major disadvantage of the compromise is that international standardisation is abandoned, with equipment and airplane manufacturers developing the multimode receivers needed, and looking at several different ways of improving basic GNSS derived signals so as to use them for the lowest operational visibilities with which ILS/MLS based systems can contend and/or using totally different landing systems.

GNSS based landing systems are termed Global Landing Systems (GLS), and satisfactory automatic landings have been performed using these systems in good weather, but a system of signal refinement and/or augmentation is necessary to equal ILS/MLS. One proposed system will provide 35 ground reference stations providing 'wide area' augmentation over the whole of the USA to produce an operating standard equal to Category 1 ILS/MLS. A second proposal requires a locally based 'differential' transmitter station that measures to an accuracy better than one meter, the errors and anomalies of GNSS signals, and transmits ILS-like signals to the aircraft by data link, but having the disadvantage that each installation would need to be certificated and user aircraft carry a data base of 'pseudo' approaches. This may permit service to *all* runways within a 30 mile radius of the 'differential' transmitter so providing greatly reduced costs compared with ILS/MLS. Protagonists of the systems are confident that GLS will eventually perform as well as ILS and MLS. With *all* of these proposals, airlines require that no changes be made to the *aircraft's* landing systems to use the new technologies, other than provision of data link. It seems likely that Europe with its greater need for so-called 'all weather operations' will not wait for these longer term developments and will use ILS and MLS into the 21st century.

Lockheed-Martin has proposed an Autonomous Precision Approach and Landing System (APALS) using GNSS to update an inertial reference unit for initial approach, and then the system compares aircraft position with a terrain data base for precision approaches. The system can generate an ILS look-alike data stream to allow automatic landing, or act as prime sensor for a head up display (HUD). A key process in APALS is the enhancement of an aircraft's weather radar to produce ground images of sufficient resolution to allow matching with a terrain data base. It is claimed that Category 3b sub-meter navigational accuracy is achievable, based on experience with the Pershing 2 ballistic missile that used similar techniques. The system is independent of ground based aids.

Head up displays (HUD)

Airlines and equipment manufacturers are looking at other means of improving regularity of operations, with some examining the possibility of using head up displays (HUD) to augment visual guidance from approach and runway lighting systems and radio guidance from conventional landing systems. The simplest form of HUD projects on to the windshield the same information displayed on conventional flight instruments but focused at infinity. The important advantage is that a pilot is looking through the display, and can see approach lights and runway at the earliest moment, and carry out a manual landing or monitor an automatic landing without the difficult task of adjusting from head-down to head-up immediately before the landing.

More advanced HUDs augment basic information by adding computer derived 'pictures' of the runway, a flightpath vector, aircraft total energy symbols and projected touch-down point. The additional sensors required may be from accelerometers or inertial navigation systems (INS).

These systems are increasingly used in the USA and Europe, and the French airline Air Inter adopted HUDs as early as 1974, and continues to use them on all airplane types even where automatic landing systems are installed. In the USA, Alaska Airlines used a Head up Guidance System (HGS) on its airplanes by 1984, and achieved *manual* landing minima of 50 feet decision height and 210 meters runway visual range

(RVR), a remarkable achievement. In the take off case, minimum RVR is reduced from 210m to 90m, and Alaska Airlines states that unit cost of HGS is $300 000 and that maintenance costs only 8% of those for autopilot and autothrottle systems. The Flight Safety Foundation believes that of 1000 accidents studied, a HUD might have prevented more than 31%, and the FAA has stated that a hybrid system comprising a fail passive autoland system, and a fail passive HUD could achieve Category 3b – zero decision height and RVR of 90 m.

More advanced, but not yet operational, HUD systems are Enhanced Vision Systems (EVS), and they may use digital terrain data that matches terrain profile measured by radar altimeters against database profiles, or Forward Looking Infra Red (FLIR) or millimeter wave length airborne radar. Comparisons of FLIR and mm/wave radar performance show that each performs well in different weather conditions, depending on air temperature and size of water droplets in the atmosphere, and a mix of sensors may therefore be required to provide an always-available video image of a runway at up to 6 km range. It is expected that EVS will make airplanes independent of ground-based electronic aids and able to operate in restricted visibilities on more than 5000 runways at USA airfields. Its performance characteristics provide additional benefits such as detection of ground obstacles, so aiding taxying in restricted visibility. Northwest Airlines has a fleet of airplanes that does not justify the expenditure required to provide autoland systems, and is actively pursuing (in 1996) acquisition of EVS, as are other major USA airlines.

An ultimate form of EVS might be to provide Synthetic Vision Systems (SVS) using a terrain data base and sensors for back up, eliminating the need to see the 'real' world, a concept that may attract designers of future supersonic airliners wishing to avoid the complicated and heavy nose droop system of Concorde.

Conclusion

A book of this kind is never complete for the rate of growth and technological change of the civil air transport industry produce new safety challenges for aircraft designers, constructors, airlines, regulatory authorities and pilots. New aircraft, engines, systems, instruments, increased automation, computers with and without artificial intelligence, new materials, a changing operational and economic environment, and the frailty of man will ensure that incidents and accidents continue to happen and as one safety problem is identified and solved another will appear.

Continued growth may produce two billion passengers per year by the end of the century, and if the accident rate equals that of the *best* year in aviation history, a thousand passengers will lose their lives each year. Pressures to reduce fares will continue and produce challenges to the regulatory authorities, aircraft and engine manufacturers and airlines. Many accepted practices affecting safety will be challenged, including aircraft and engine maintenance and overhaul schedules and safe-life and fail-safe concepts for structures. Unfamiliar problems will arise from the use of new materials, for even the best research and development programs fail to foresee all possible circumstances and failure modes. Another unknown is whether crimes of violence will continue to put lives at risk, and whether airport security systems can cope with two billion passengers per year, if there is an upsurge in terrorism.

It is hoped that those in positions of responsibility and influence will ensure that civil aviation continues to progress towards that unattainable ideal, perfect safety, and that risk management techniques evolved will be taken up by other modes of transport, some of which have a poor record in comparison with air transport! Safety organisations and programs needing additional resources if air safety is to be improved, or even maintained at recent levels, include the following:

■ Accident investigation, perhaps to include the funding of an International Bureau of Aircraft Accident Investigation to ensure that establishing cause and not assigning blame has top priority

- Separation of safety from other regulatory programs of states
- Increased activity and funding for ICAO, and particularly for its technical cooperation programs
- Early global implementation of improvements to the operational environment, with satellite navigation and control systems introduced to provide an acceptably safe environment for increased air traffic in airspace throughout the world, backed up by ACAS systems
- More airports and runways to reduce air traffic congestion threatening safety
- Better matching of airports and operational tasks, particularly with regard to aerodrome ground aids and emergency equipment
- Improved packing, inspection and surveillance systems for dangerous goods carried on aircraft and elimination of the need for 'combi-aircraft'
- Improved training for flight and cabin crews, particularly in emergency procedures
- Improved design criteria and safety equipment for aircraft cabins to make them safer in emergency conditions and less difficult to evacuate in adverse conditions
- More research into human factors and particularly into the interfaces and relative tasking of man/computer/machine
- Improved techniques for airport security systems and better training for operators of the systems
- Improved safety information exchange systems
- An equalisation of risk, with airlines with the 'worst' records in air safety being improved to the standards of the 'best'

The cost of funding these improvements is not calculable except by safety authorities, airlines and manufacturers. Improved operating efficiencies of the latest aircraft and a continuing availability of cheap fuel should produce reduced costs in real terms and enable required safety programs to be commissioned without greatly affecting passenger fares. A major problem is to ensure that improvements are introduced on an international and coordinated basis, so as to avoid what airlines see as competitive disadvantages. If the improvements added as much as 2% to the cost of an airline ticket, it is difficult to believe that growth in air travel would be significantly affected, and that the increases would not be accepted by the travelling public if explanations were provided. If a concerted and international safety program is implemented it would be reasonable for the total number of accidents and fatalities to remain constant in an industry expected to achieve a growth rate of 7% per annum. This achievement would make air travel safer than competing travel modes and ensure further growth in the future.

The answer to the question, Air Travel: How Safe Is It? must therefore be that it is safe in 1996 and will become safer each year if resources are committed to a coordinated and globally implemented air safety program!

Index